CATTLE FEEDING

CATTLE FEEDING

JOHN B. OWEN
MA(Cantab), BSc, PhD (Wales), FIBiol, FRAgS
Professor of Agriculture, University of Wales, Bangor

FARMING PRESS

First published 1983
2nd edition 1991

Copyright © John B. Owen, 1983, 1991

All rights reserved. No part of this publication may be reproduced, stored in a retrieval system, or transmitted, in any form or by any means electronic, mechanical, photocopying, recording or otherwise, without the prior permission of Farming Press Limited.

British Library Cataloguing in Publication Data

Owen, John B. (John Bryn)
Cattle feeding.
1. Dairy cattle—Feeding and feeds
I. Title
636.2′142 SF203

ISBN 0 85236 220 X

Published by Farming Press Books
4 Friars Courtyard, 30–32 Princes Street
Ipswich IP1 1RJ, United Kingdom

Distributed in North America
by Diamond Farm Enterprises,
Box 537, Alexandria Bay, NY 13607, USA

Printed in Great Britain by Butler and Tanner, Frome, Somerset

Contents

Illustrations

PLATES

FIGURES

TABLES

Page

PREFACE

THE ART of cattle feeding is changing and developing all the time. Some basic fundamentals remain the same but with new knowledge, new feed materials and changing economic circumstances, systems of feeding are undergoing marked development.

In the past there could be quite a gulf between the theoretical nutritionist and the feeder at the sharp end of the daily responsibility. To make the most efficient use of resources nowadays this gulf must be bridged so that communication between theorist and practitioner is improved and so that the enterprising cattle feeder can reduce lag in putting current research into feeding practice.

The aim of this book is twofold. In the first place it is intended to present a reasonably clear picture of how current knowledge of the nutrition of cattle is incorporated into cattle feeding practices. Secondly an attempt is made to combine nutritional knowledge with economic information to answer the crucial question of how to feed cattle efficiently and profitably. The careful reader of this book, after having checked on the price he expects for his product, the cost of his concentrate feed and having estimated the cost of his forage, should have a reasonable estimate of the economically optimum diet to feed his cattle. Time will judge whether this claim is justified but few previous books on cattle feeding have even attempted the task and most do not mention money in the context of cattle feeding.

The book therefore should prove valuable for cattle farmers and for those who serve them as advisors and suppliers of their various needs. It is also intended as a textbook on cattle feeding practice suitable for degree and diploma level students in Agriculture and Veterinary Science.

Colleagues in research may care to consider the many gaps in knowledge that have become apparent in trying to put forward a coherent practical package from the various threads of current advances in knowledge. If this were to stimulate some further research to strengthen the basis of current feeding practices, then the book would have achieved its purpose.

I am indebted to many colleagues who have helped in the writing of this book. In particular I am especially grateful to Mr C. T. Livesey M.Sc., B.V.Sc., MRCVS for his enthusiastic and unstinting collaboration over the last few years and in the preparation of Chapter 9. I am also much indebted to Dr A. G. Chamberlain B.Sc., Ph.D. and Mr. O. Kroll M.Sc., for their collaboration in cow experiments at Bangor and to Mr R. T. Williams the Manager of the University Farm whose sound practical guidance has been invaluable. Lastly I wish to acknowledge the technical and secretarial assistance in the final production ably co-ordinated by Mrs. P. Ellis.

Bangor
September 1983

JOHN B. OWEN

Chapter 1

CATTLE PRODUCTION SYSTEMS

As a background to the study of feeding it is important to consider the systems of cattle production to which it relates. These systems include milk production and beef production although most cattle systems involve the joint production of both.

CATTLE ARE widely spread throughout the world wherever agricultural production is possible and have a more even distribution than sheep and goats, although they are not so commonly found in the extremely hostile environments (Table 1.1). In most developed parts of the world, cattle are the main source of milk. However, there are parts of the world where sheep are the main dairy animals and others, such as Egypt, where the buffalo is the main source of milk (Table 1.2).

In India, cattle-keeping is markedly influenced because of the religious protection of the cattle beast. In India therefore cattle are used mainly for dairy, draught and ceremonial purposes.

When cattle are used for dairy purposes they are normally also the source of beef production whether this be the meat of young calves (veal) or that of older cattle (beef) (Table 1.3). Calves may either stay on the farm of their birth until they reach slaughter stage or they may be sold as calves or as store cattle at various stages, to be finished by other producers.

Specialist beef production is not so widespread and is found mainly in North and South America (such as Argentina) and in other parts of the world in the margin between temperate arable production and mountainous lands (as in the United Kingdom) or in semi-arid areas as in Australia.

Many of the calves produced from the specialist beef herd,

Table 1.1. Human and animal populations of some important cattle countries

Continent or Country	*Land Area (m.ha)*	*Human Popn (m)*	*Popn. in agr %*	*Cattle No. (m)*	*Goat No. (m)*	*Sheep No. (m)*	*Cattle per 100 ha*
Africa	*2964*	*610*	*64*	*181*	*167*	*200*	*6*
Ethiopia	110	45	76	31	17	23	28
Kenya	57	23	78	10	9	7	18
Nigeria	91	105	66	12	26	13	13
South Africa	122	34	15	12	6	30	10
Sudan	238	24	63	23	14	19	10
N & C America	*2138*	*417*	*11*	*164*	*15*	*19*	*8*
Canada	92	26	3.6	12	0.03	0.7	13
Cuba	11	10	20	5	0.11	0.4	45
Mexico	191	85	31	31	10	6	16
USA	917	246	2.5	99	2	11	11
S. America	*1753*	*285*	*24*	*257*	*22*	*111*	*15*
Argentina	274	32	11	51	3	29	19
Brazil	846	144	26	134	11	20	16
Uruguay	17	3	14	10	0.01	26	59
Venezuela	88	19	12	13	1.4	0.4	15
Asia	*2769*	*2994*	*61*	*384*	*296*	*332*	*14*
Bangladesh	13	110	70	23	11	1.1	177
China	933	1101	69	74	78	103	8
India	297	819	67	193	105	52	65
Israel	2	4	5	0.32	0.13	0.28	16
Japan	38	123	7	5	0.04	0.03	13
Pakistan	77	115	51	17	33	27	22
Sri Lanka	6	17	52	1.8	0.5	0.03	30
Turkey	77	54	50	12	13	40	16
Europe	*473*	*497*	*10*	*125*	*12*	*142*	*26*
Austria	8	8	6	2.6	0.03	0.02	33
Bulgaria	11	9	13	1.6	0.4	9	15
Czechoslovakia	13	16	10	5	0.05	1.1	38
Denmark	4	5	5	2.3	—	0.13	58
France	55	56	6	21	1	10	38
Germany	35	78	5	21	0.1	4	60
Ireland	7	3.6	14	5.6	0.01	4	80
Italy	29	57	8	9	1.2	12	31
Netherlands	3.4	15	4	4.5	0.3	1	132
Poland	30	38	22	10	0.01	4	33
Romania	23	23	22	7	1	19	30
Spain	50	39	12	5	3	18	10
UK	24	57	2.1	12	0.06	28	50
Wales	1.7	2.9	1.9	1.3	—	10.3	76
Oceania	*843*	*26*	*17*	*32*	*2*	*229*	*4*
Australia	762	16	5	23	0.003	164	3
New Zealand	27	3	9	8	1.3	65	30
USSR	*2227*	*286*	*14*	*121*	*6*	*141*	*5*
World	*13077*	*5115*	*47*	*1264*	*520*	*1173*	*10*

Source: FAO *Production Yearbook 42*, 1989.

Table 1.2. Liquid milk production from various sources in 1988

	Total Production 1000 mt	*% from various sources*			
		Cows	*Buffalo*	*Sheep*	*Goats*
Africa	17445	72	8	9	11
N & C America	85208	100	—	—	—
S America	29311	99	—	—	1
Asia	93015	52	40	4	4
Europe	172594	100	—	—	—
Oceania	14209	100	—	—	—
USSR	106396	100	—	—	—
World	518178	89	7	2	2

Source: FAO *Production Yearbook 42*, 1989.

Table 1.3. Beef in relation to other meat production

	Total Meat production 1000 mt	*% from various sources*				
		Cattle	*Sheep*	*Goat*	*Pig*	*Poultry*
Africa	8267	39	10	7	6	24
N & C America	34817	39	0.6	0.1	26	33
S America	12188	57	2	0.6	16	28
Asia	41955	8	5	4	59	20
Europe	42861	25	3	0.2	50	19
Oceania	4239	51	28	—	9	11
USSR	19213	45	4	1	33	16
World	163540	30	9	1	39	23

Source: FAO *Production Yearbook 42*, 1989.

after weaning from the cow, may join the pool of store cattle from various sources.

The final stage of feeding beef cattle to slaughter may take place in a variety of units but this type of production is amenable to economies of scale and large feed lots may be found in some countries where cattle are finished on large mechanised units similar to that shown in Plate 1.

DAIRY CATTLE SYSTEMS

Dairy cattle systems vary from the very small herds associated with peasant farming (Plate 2) to large herds where several

Plate 1. (*above*) A feed lot in the United States (*Dairy Farmer*)

Plate 2. (*below*) Dairy farming near the Bavarian Alps (*A. I. D., Boun*)

hundred cows may be kept in one herd. However it is the medium-sized family unit that is the backbone of present-day dairying both in America, Western Europe and many of the other dairying areas.

Table 1.4 shows the size of dairy herds in the 12 EC countries. It is during the present century that dairying has developed from a subsistence activity through the sale of butter and cheese made on the farm to augment money income, to a commercial activity where all the milk produced on a farm is sold in the liquid form to large milk marketing bodies. Most of the milk produced is processed in creameries or factories into milk products to be sold to retail outlets, although a proportion is sold as milk for liquid consumption. In the UK, where approximately 50 per cent of the milk produced in 1982 was sold as liquid milk, there is a much higher proportion of liquid sales than in most other countries.

Feeding Systems

The variation in the type of dairy system and in particular the variation in size of herd has major repercussions on other aspects such as the housing system and this in turn influences the type of feeding system. Table 1.5 also shows the type of

Table 1.4. Average size of dairy herd in 12 EC countries

Country	Average herd size
Germany	15.3
France	19.9
Italy	9.1
Netherlands	41.5
Belgium	23.3
Luxembourg	30.8
United Kingdom	60.0
Irish Republic	19.9
Denmark	28.2
Greece	3.0
Spain	6.1
Portugal	3.2

Source: *EC Dairy Facts and Figures* 1987.

Table 1.5. Housing and feeding systems for dairy cows in selected European countries

	HOUSING (% of all herds)				*FEEDING (% of all herds)*		
	Cow stalls	*Yard & parlour*	*Cubicles & parlour*	*Housed all year*	*Silage & concentrates*	*Silage hay, roots & concentrates*	*Hay & concentrates*
Belgium	—	NA	—	—	70	30	—
Denmark	98	2	—	1	10	70	20
W. Germany	—	NA	—	—	—	60	40
Luxembourg	96	—	4	—	—	45	55
Netherlands	84	—	15	1	30	65	5
Norway	97	1	2	—	80	15	5
Scotland	68	32		—	20	14	66
N. Ireland	58	1	41	—	53	—	47
England & Wales	46	24	30	—	58	2	40

Source: Owen, J. B. (1987) *In* Dairy Cattle Production, H. Gravert (Ed.), Elsevier Science Publishers, Amsterdam, Ch. 3.

housing and feeding system associated with dairying in the various countries of Europe.

Dairying is mainly concentrated in temperate areas of the world with a distinct seasonal pattern of winter and summer although the extent of the seasonality varies considerably.

At one extreme New Zealand is considered to have one of the most suitable climates for dairy production, with a mild winter and a reasonably even rainfall which maintains grass growth over a long growing period (Plate 3). At the other extreme some of the northern European countries have a long severe winter with extensive snow cover and a shorter but intensive period of summer growth. At the lower end of the temperate latitude or inland from the temperate coastal areas, drought becomes an increasing problem which can be seen at its most intense in the developing dairy industries of the Middle East.

Because of this range of climatic conditions under which dairying is carried out there is also a distinct gradation in type of feeding system. At the one extreme cattle are housed for more than half the year and fed on conserved forage supplemented with grain-based concentrates. During the

Plate 3. (*above*) A dairy farm in South Auckland, New Zealand
(N.Z. High Commission)

Plate 4. (*below*) Welsh Blacks—the premier British beef suckler breed at the Haulfryn group breeding scheme (Mr E. Pritchard, UCNW)

grass growing season the cows graze on pastures without, or with very limited, supplementation. At the other extreme, where pasture growth is not dependable, cattle may not graze on pasture at any time, all the feed being given in the form of conserved or fresh cut forage, often grown under irrigation. Such systems are common in Israel and are becoming a feature of other countries in the semi-arid zone.

All of these variations in management systems markedly influence the nature of the problem of feeding cattle. For example, formulating supplements to grazing is a different exercise from that of designing diets to be given as the sole source of food for the herd.

The nature of the forage and grains available affect the type of feeding possible. In some of the areas of lower latitude, maize forms the staple forage for cattle feeding, both in the form of whole maize silage, maize grain and of maize cobs. There are thus major regional differences in feeding that arise from many aspects.

Autumn versus Spring

A major factor that can influence the feeding and management of a dairy herd is the pattern of calving over the season. Traditionally dairy farmers producing milk for manufacture into butter and cheese have tended to adopt a policy of spring calving where the cows calve just before turning out to pasture in the spring and produce their milk off summer grass without grain supplementation.

On the other hand in countries like Britain, where there is a larged demand for liquid milk, the production pattern necessitates that a proportion of cows calve in the autumn and early winter. These cows are in full production during the period of indoor feeding and are therefore far more dependent on grain supplements than the spring-calving herd (see Table 1.6).

Beef Cattle Systems

Specialist beef production, where cows are kept solely to suckle their calves, is almost entirely confined to range land areas where the cows can make use of extensive grazing to

Table 1.6. Some features of a sample of British dairy herds classified according to seasonality of production

	Autumn-calving herds (<46% of milk produced in April–Sept. inclusive)	*Spring-calving herds (>60%)*
No. of herds	255	68
Herd size	113	93
Yield/cow (litres)	5,393	4,562
Concentrate use per cow (kg)	1,842	1,415
Stocking rate (LSU/ha)	2.01	1.89

Source: Milk Marketing Board, *An analysis of FMS costed farms 1981–82,* Report No. 33.

suckle their calves during the main grass-growing period. These beef cattle herds may be large, such as those seen on films depicting the life of the early United States and still a feature of the Rocky Mountain plain areas, or they may be small such as those associated with sheep production in the upland areas of the United Kingdom (Plate 4).

These cows may be wintered outdoors with forage supplementation or, in the wetter and colder areas, kept indoors.

As with dairy herds, seasonality of production can markedly affect the management and feeding procedures. Traditionally many beef cattle herds would calve in the spring just before the most active period of grass growth so that the major part of lactation coincided with the grass-growing season. In the autumn, calves would be weaned and either sold or treated separately and the dry cows would be kept over winter.

More recently calving date has tended to become earlier and earlier so that in the United Kingdom, for instance, many beef herds now calve in the late summer and autumn. This radically changes the pattern and quality of feed input but results in calves that can be weaned in the spring and early summer and sold as much larger cattle before the winter, as shown in Table 1.7.

Table 1.7. Some physical performance data for upland beef herds in the U.K. according to season of calving spring (March, April, May) and autumn (Sept., Oct., Nov.)

	Spring	*Autumn*
Calves reared (%)	88	97
Calf age at weaning (days)	209	303
Weaning wt. (kg)	231	293
Daily gain	0.92	0.83
Cow concentrates (kg)	92	132
Calf concentrates (kg)	35	135
Silage (tonnes)	2.2	4.9
Hay (tonnes)	0.7	0.3
Other feed & straw (tonnes)	0.5	0.5

Source: MLC (1982), Beef improvement services Data Sheet 82/4.

Trends in Cattle Production Systems

Over several decades there has been a trend in most countries for the size of herd to increase. This is true of both dairy herds and beef herds. Although the number of really large dairy herds (200+) is still relatively small, there has been a steady increase in the size of herd kept by one family (Fig. 1.1). This stems from the fact that the easiest way to improve profitability is to expand the herd if this can be done without a major increase in overhead capital costs.

The EC introduced milk production quotas in 1984 to control European milk output. A similar system has operated in Canada for some time. Quotas have had a marked effect on the milk production pattern both nationally and for the individual producer.

Flexibility in terms of the sale and leasing of quotas has however still enabled some of the trends in the rationalisation of dairy herd size to continue.

This trend to larger herds has been associated with a change from the traditional cow byre or cow stall system, where each cow is kept in a stall for feeding, milking and living, to systems where milking has been separated from

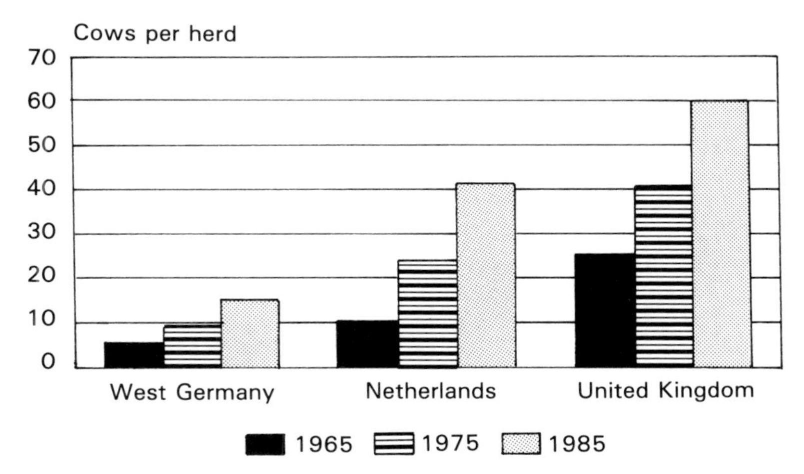

Figure 1.1. Dairy herd size: trend from 1965 to 1985
Source: EC Dairy Facts and Figures

the other activities. This system of a specialised milking parlour, associated with separate areas for feeding forage and for lying, has facilitated progressive expansion in dairy herd size since the capacity of the milking parlour is not the factor usually limiting herd size. The provision of extra lying space for the cow is often fairly easily accomplished.

So far the trend to increased cow herd size has not resulted in large numbers of herds where 200 or more cows are kept as one herd. This may be due to the fact that attention to detail in milking and feeing is still better accomplished by having a system where one man carries the responsibility and is intimately involved in all the work that goes on.

With beef cattle the situation is rather different in that very large herds of beef cows have been a feature of many range-land beef units and, more recently, very large-scale feed lots have been common in the United States. Economies of scale are much more evident here and since milking is not involved, there is little difficulty, in areas where the climate is relatively dry, in keeping thousands of finishing beef cattle in outdoor paddocks.

In wetter areas the necessity to provide roofed areas considerably increases capital costs and restricts the expansion of the size of the finishing unit.

References

BROSTER, W. H. and SWAN, H. (1979), *Feeding Strategy for the High-Yielding Dairy Cow*, Granada Publishing Limited.

FAO (1989), *Production Yearbook 42.*

M.M.B. (1982), Production Division Report.

PHILLIPS, C. J. C. (ed.) (1989), *New Techniques in Cattle Production*, Butterworths, London.

PRESTON, T. R. and WILLIS, M. B. (1975), *Intensive Beef Production*, Pergamon Press.

Chapter 2

PRINCIPLES OF CATTLE FEEDING

Understanding is the foundation of innovation and flexibility; this is particularly so for feeding practice, where there are continuing developments which reduce the value of rigid standard recipes and procedures.

CATTLE ARE ruminants, with very specialised digestive systems, adapted particularly to utilise a range of forages of varying quality.

The calf when born is similar to the young of other animals in that the milk it takes in passes down the gullet or oesophagus and into the true stomach or abomasum. The rumen is at this stage small and undeveloped and completely by-passed by liquid sucked by the calf. As the calf grows and nibbles solid food, this enters the rumen which develops very rapidly from the age of two to three weeks if milk is restricted. The rumen in the adult is a large sac-like organ containing about three-quarters of the gut contents which may be up to 100 kg in total.

Before the calf is weaned, milk or milk replacers in the liquid form cause the calf to go into a sucking mode. The outward sign of this is the excitable behaviour accompanied by vigorous tail wagging and butting, particularly if the milk flow is restricted. Inwardly the rumen wall forms a tube-like groove which acts as an extension of the oesophagus and is called the oesophageal groove. The function of this is to by-pass the rumen, a most important necessity for the well-being of the young calf. The calf can go into a sucking mode even when it has been trained to drink its feed from a bucket, and the 'drinking' of milk from a pail is a form of sublimated sucking quite distinct from the drinking of water. Occasionally under some circumstances the oesophageal groove reflex does not function properly and calves become unthrifty and ill.

In the first 3–4 weeks of life the calf, like many other young animals, has a very restricted capacity to digest feed material because of the rather simple and restricted range of enzymes it can produce to metabolise the food. Cow's milk is the ideal food for the young calf and there is only limited room for manoeuvre in trying to formulate cheaper materials to replace milk or milk products for the first few weeks of the calf's life. The chief constituents of cow's milk are the protein, carbohydrate and fat and each of these presents problems when cheaper replacements are sought. These are discussed in more detail in Chapter 6 but one feature deserves special mention. Much of the protein in cow's milk is in the form of casein which has the peculiar property of forming a curd or clot under the action of the rennet and hydrochloric acid secreted in the calf's abomasum. When a calf has had a drink of milk the milk forms into one large curd in the abomasum which entraps much of the fat as well as the protein. This has the important function of holding the milk in the abomasum whilst it is gradually released, by the breaking down of the curd through the action of the digestive enzymes. A reasonably uniform release of nutrients over the twenty-four hours is thereby ensured rather than the spasmodic flushes of milk that would otherwise occur. This enables the calf's digestive and metabolic systems to cope with their complex tasks. It is apparent that replacing casein with other proteins is a much more difficult task when the significance of the clotting mechanism is appreciated.

DIGESTION IN THE RUMINANT

The main elements of the digestion of food in cattle are the mouth with its teeth, tongue and lip, the rumen with its microbes, the abomasum and the small intestine. Food material enters the mouth under the joint manipulation of the tongue, the lip and the incisor teeth on the lower jaw biting against the hard toothless pad of the upper jaw.

Some preliminary chewing between the two sets of grinding

molars takes place as well as the admixture of saliva. This food material then passes down the gullet into the rumen, which can be likened to a large fermentation vat containing a mixture of fluid and food material. The rumen contents are the substrate for a large population of micro-organisms, both bacteria and protozoa.

Several types of microbes make up the microbial population of the rumen and the constitution of this population—the microbial flora—depends on many factors, the chief being the diet of the animal. Some of the bacteria—the cellulolytic strains—can break down some of the tough structural elements of the plant and make the plant nutrients available to themselves and eventually to the host animal. In the process of anaerobic fermentation that goes on in the rumen, protein and other nitrogen compounds are broken down into ammonia and, under appropriate conditions, much of this can be used by the microbes to synthesise their own body protein. Also in the rumen some of the essential vitamins can be synthesised by the microbes, e.g. vitamin B_{12}, provided the necessary cobalt ingredient is present in the diet.

The result of the fermentation is the breakdown of food material, including cellulose, into volatile fatty acids—primarily acetic, propionic and butyric acids—which are absorbed through the rumen wall into the bloodstream and form the main source of energy for the ruminant. Some of the excess ammonia may also be absorbed and excreted as urea. Another by-product of the fermentation process is methane gas which the beast regularly gets rid of by the process of belching.

As well as fresh food material being regularly added to this large mass of fermenting rumen contents, the more fluid fraction, containing the smallest particles, regularly flows out of the rumen via the ruminal–omasal orifice and passes on to the abomasum. This material contains the bodies of micro-organisms, which are a rich source of protein and undegraded food material, which has escaped fermentation in the rumen.

One other added feature of rumen digestion is rumination, the process whereby boluses of coarse food material are formed and regurgitated for a further mastication in the mouth before swallowing again.

In the abomasum and intestines the digestion process common in non-ruminants takes place, resulting in the absorption, through the wall of the small intestine, of amino acids, carbohydrates and the breakdown products of fats or lipids.

Several features of the process of ruminant digestion are important in understanding the basic principles of cattle feeding.

1. Protein Supply

Because of its micro-organisms the cattle beast can survive and in certain cases, thrive, on a protein-free diet, provided it contains a source of nitrogen compounds as a substrate for protein synthesis. The classic experiments of Virtanen in Finland showed that cows could live and produce milk on a protein-free diet (Table 2.1). However the same experiments showed that cows cannot produce high or normal yields on such diets. Young growing cattle and milking cows have such a relatively high need for protein that the maximum production of microbial protein does not supply all their amino acid needs. These have to be supplemented by good-quality protein which by-passes the rumen and is absorbed in the small intestine. The calf can obtain this protein in the pre-weaning stage by the action of the oesophageal groove which mechanically by-passes the rumen

Table 2.1. A comparison of milk production of cows on protein-free diets with those on normal diets containing protein

	Protein-free feed with urea and ammonium salts	*Normal diet containing true protein*
No. of cows	15	19
Milk production (kg)	2,945	5,078
Butterfat (%)	5.1	4.0

Source: Virtanen, A. I., Ettala, T. and Mäkinen, S. (1972), in 'Festskrift Til Knut Brienem pp. 249–76, Utigitt av en Redaksjons komite.

if liquid feed is given. This process can be continued into adulthood if the beast is not weaned. However once weaned, adult cattle depend on the presence in their diet of protein that is not degradable in the rumen—undegraded protein (UDP). This undegradability may be simply due to the natural properties of the food or because the food material has been subjected to processes which 'protect' the protein. Naturally if these processes are too effective they defeat their own object and make the protein entirely indigestible and useless! Certain heating processes can aid protein protection and chemical processing using formaldehyde is also effective. Another process known to protect proteins is a process akin to tanning although it is difficult to stop the process going too far.

Table 2.2 shows the results of an experiment where formaldehyde treatment is assessed.

Table 2.2. The effect of treating whole-crop maize silage with different levels of formaldehyde on the performance of Friesian steer calves

	Formaldehyde applied at ensiling (g/kg CP)			
	0	*14*	*27*	*53*
Daily intake of silage DM (g/kg liveweight)	22.3	24.0	24.9	25.0
Daily liveweight gain (g/day)	457	594	653	683

Source: Kaiser, A. G., Osbourn, D. F., England, P. and Dhanoa, M. S. (1982), *Animal Production 34*, 179–90.

Another way in which the use of protein can be optimised is to choose the correct mixture of feed ingredients. Feeds vary considerably in their crude protein content and in the level of degradability of this crude protein as shown later (Table 4.2).

2. The Role of Roughage

The ruminant animal has been evolved to cope with a diet of fibrous forage, so much so that it does not use low-fibre diets efficiently. Adding concentrates to a forage diet usually improves production, e.g. liveweight gain, but only up to a point.

So far it has been impossible to define the 'roughage' characteristics of a diet in a single unambiguous figure. For example crude fibre content, as normally measured, does not take into account the complicating effect of particle size. A high-fibre diet, when finely ground, loses many of the characteristics identified with 'roughage' and may have effects in the animal strikingly different from the same diet given in the unground form (Table 2.3).

Table 2.3. The effect of grinding and pelleting of two diets on DM intake and DM digestibility when given to cattle

	Unground long form		*Ground, pelleted form*	
	Daily DM intake (g/kg $W^{0.75}$)	*DM digestibility (%)*	*Daily DM intake (g/kg $W^{0.75}$)*	*DM digestibility (%)*
High-quality dried grass	84.5	71.7	85.5	55.4
Medium-quality dried grass	73.8	68.2	93.9	53.3

Source: Greenhalgh, J. F. D. and Reid, G. W. (1973). *Animal Production 16*, 223–33.

Before the effect of roughage on digestion and performance in cattle is examined the concept of roughage needs to be amplified and clarified. This can best be done at present by listing a number of attributes that are together identified with differences in roughage characteristics as follows: crude fibre, energy requirement for grinding (fibrosity), volume density, digestible energy content, cellulose/hemicellulose content, particle size, fruit/grain fraction of plants, leaf stem ratio.

Effect on cattle. Roughage has several marked effects on digestion and performance in cattle.

- *Physical effect.* Roughage seems to have a direct physical contact effect on the rumen wall; this seems to be necessary to maintain normal muscular movement in the rumen wall and to keep the roughly corrugated papillated surface intact and functioning.

- *Chemical effect.* Roughage content is associated with the composition of the rumen fluid; high roughage being associated with a higher ratio of acetic to propionic acid and lower lactic acid level. Rumen fluid is less acid, i.e. has a higher pH, when dietary roughage level is high.

High-roughage diets are in some senses 'safe' diets for cattle, for instance there is less danger of high levels of lactic acid building up when cattle are suddenly exposed to them, thus reducing the possibility of acidosis or laminitis. They are also diets that safeguard the process of milk synthesis. Whilst milk production may be depressed on high-roughage diets, because of the energy restriction imposed, in other respects milk synthesis and the composition of the milk is unimpaired although protein level can be reduced. At low levels of roughage, on the other hand, lactation becomes distinctly abnormal with the possibility of very marked reduction in fat synthesis.

End-Products of Digestion on Different Diets

Different diets can markedly influence cattle production because of their inherent properties as consumed and the way they are processed in the animal to end up as nutrients, absorbed for body metabolism. An extreme example would be the feeding of milk protein or casein to a six-week-old calf. If this were incorporated into a solid calf-starter diet, much of its value as a protein would be lost due to its being highly degradable and broken down by the rumen microbes into ammonia. Because casein is so rapidly broken down in the rumen, much of the ammonia can escape the protein-synthesising action of the microbes and be largely excreted as urea. On the other hand, if the casein was incorporated into a liquid milk replacer it would shoot straight into the abomasum and be normally utilised to provide the amino-acid needs of the calf.

Apart from the end-products from the dietary nitrogen *per*

se the main end-products of digestion that supply the energy for cattle are the volatile fatty acids, already mentioned and there is a marked systematic effect on these according to the type of diet, in particular the forage:concentrate ratio of the diet. Some of the extreme variations have just been mentioned. Table 2.4 shows the composition of the end-products of digestion in cattle in a range of diets.

Table 2.4. End-products of digestion in the rumen liquor of cattle on a range of diets (expressed on a molar basis)

	Acetic	*Propionic*	*Butyric*	*Other*
Arable silage	74	17	7	2
Poor hay	69	18	10	3
Hay & concentrates	60	23	14	3
Alfalfa hay & grain	65	21	11	3
Ground hay & maize	39	38	10	13

Source: Blaxter, K. L. (1962), *Energy Metabolism of Ruminants*, Hutchinson.

In general increasing the forage:concentrate ratio results in progressive increases in acetic, falls in propionic and less clear trends in butyric acid. These acids are absorbed across the rumen wall into the bloodstream and form the substrate for many of the life processes of cattle.

Conversion of Nutrients into Various Outlets in Cattle

The process of concern can be generally summarised as shown in Fig. 2.1.

The level of intake of food and the type of food profoundly influence the whole process illustrated in Fig. 2.1 which can be taken to refer to any cattle beast, including the milking cow.

When the level of intake is restricted it can readily be seen from Fig. 2.1 that the flow of energy into the production process is markedly diminished. In the milking cow this can result in some reduction in milk output but its major consequence is an increased partition of the diminished net energy into milk at the expense of that going into body tissue.

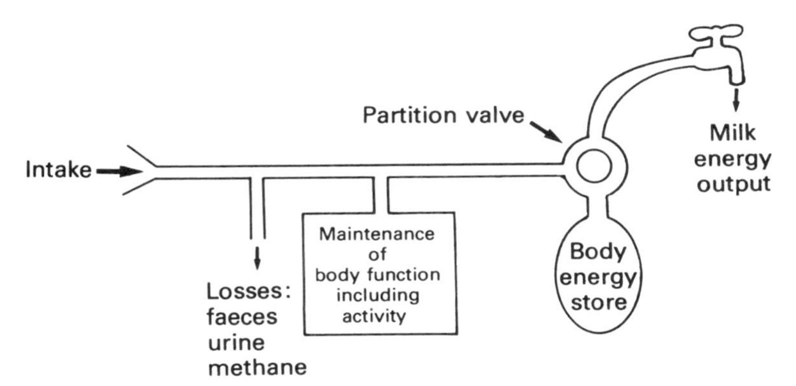

Figure 2.1. Partition of feed energy in the cow

Often this can result in a complete reversal from a gain in body tissue to a loss of body tissue.

In the beef animal, the restriction of energy intake reduces bodyweight gain and can affect the differential partition of the energy into the major tissues, e.g. fat, muscle and bone.

Another interesting aspect of the effect of diet on conversion into various products is the effect of the quality of the diet (Table 2.5). Some of these effects have already been referred to in relation to the volatile fatty acids.

Table 2.5. Efficiency of conversion of metabolisable energy into various functions according to type of diet

Function	*Diet metabolisability at maintenance (ME/GE)* 0.40	0.50	0.60	0.70
Maintenance	0.64	0.68	0.71	0.75
Growth & fattening	0.32	0.40	0.47	0.55
Lactation	0.56	0.60	0.63	0.66

Source: ARC (1980), *Nutrient Requirements of Ruminant Livestock*, Commonwealth Agricultural Bureaux.

Some of these main effects can be summarised as follows:

The Utilisation of Varying Volatile Fatty Acid (VFA) Mixtures

1. Dairy cattle. Many aspects of lactation efficiency have been shown to be associated with the variation in VFA ruminal pattern illustrated in Table 2.4. These include overall level of milk secretion but also the differential synthesis of the various milk constituents. Optimum lactation efficiency seems to be associated with an intermediate ratio of acetic to propionic acid. At this level the efficiency of conversion of feed into milk is maximised, butterfat synthesis is at a high level and the partition of energy into milk, at the expense of body tissue, is maximal.

To say that efficiency is associated with a certain VFA pattern does not necessarily imply a direct causal relationship, although the role of acetic acid in fat synthesis is well known and points to some causal relationship.

One of the difficulties of understanding this complex association between diet and lactation efficiency is the varying timescales over which the effects may be mediated. For example, some effects of diet changes are more or less immediate as is often seen when cows are turned out to grass in the spring. Other effects take much longer to become apparent. For example, experiments carried out at Cambridge University Farm in the 1960s compared Friesian heifers given diets varying in roughage content. In the early part of lactation, differences in milk quality, particularly milk fat content, were not striking. However by late lactation the group on the low roughage diet had very low milk fat levels, with some individuals as low as 1.5 per cent as compared to the normal level shown by the other group.

2. Beef cattle. Efficiency of beef production has also been shown to be associated with diet quality and VFA pattern (Table 2.5). The most efficient conversion into beef seems to be achieved at a lower ratio of acetic to propionic acid than that for lactation. Again causal, direct, relationships are difficult to establish since it is not possible to compare diets giving rise to different VFA patterns without changing many other things that could be the primary factors involved.

The Role of Essential Nutrients in the Conversion of Feed into Cattle Products

Apart from the major effects of energy and the substrates that provide it in the diet, other nutrients play a key role in the partition of energy and its conversion into milk or beef.

Recently, the important role of protein and the pathways by which it is provided, in the dairy cow, have been more fully realised. It is apparent that a restriction of protein supply to the cow's tissues, in terms of an optimum array of amino acids, can vitally affect the partition of energy between milk production and body gain. The moderate-yielding fat cow is the extreme manifestation of a more widespread, often unrecognised, lowering of efficiency due to incorrect diets. A fuller knowledge of protein nutrition and the protein status of feed ingredients in dietary mixtures may enable significant improvement in efficiency of milk production at no extra cost; simply by wiser deployment of the range of raw materials available.

Other essential nutrients may be shown to have similar roles in the energy metabolism of the dairy cow. Already there is evidence that the level and form of fat in the diet can affect lactation efficiency. Minerals and vitamins are also known to play important roles although it is still difficult to be sure of the appropriate dietary levels in all circumstances and types of dietary mixtures.

The beef production process may also be affected by the level of essential nutrients. Lack of protein is known to affect not only the level of intake but also the partition of energy into fat as against the non-fatty tissues. These effects are obviously more likely, in practice, to be important in the earlier stages of growth where the required ratio of other nutrients to energy is much higher. The more mature beef animal is often able to achieve optimal levels of nutrients in many practical diets without particular emphasis being placed on this aspect.

Effects of Feeding on Product Quality

Beef cattle. Beef quality as manifested in the fat:lean:bone ratio and in the distribution of meat over the carcass is influenced by genotype (breed, cross and individual) and by body weight. The level of feeding can obviously influence weight for age and can therefore influence the size and quality of carcass at a given age. The effect of level of feeding is less marked when carcasses are compared at the same weight but differences in carcass fat content have been established due to level of feeding even when carcass weight is taken into account (Table 3.6).

In addition it is also known that diet quality as well as level of feeding can influence carcass composition especially the level of fat in the carcass. Diets deficient in protein not only result in a lower growth rate and lower intake but can also increase the fat to lean ratio in the carcass.

Dairy cattle. Frequent reference is made to the effects of feeding on milk quality and it is not proposed to elaborate at this stage. Suffice to say that milk fat levels are often depressed and SNF, particularly protein content, elevated on high-concentrate diets. In the long term good feeding of dairy cattle leads to both high fat content and high SNF levels in milk as demonstrated in the Danish experiments (Table 7.4).

REFERENCES

BLAXTER, K. L. (1962), *Energy Metabolism of Ruminants,* Hutchinson & Co. Ltd.

HARESIGN, W. and COLE, D. J. A. (1981) (Editors), *Recent Developments in Ruminant Nutrition,* Butterworths.

KAUFMANN, W. (1976), 'Influence of the composition of the ration and the feeding frequency on pH regulation in the rumen and on feed intake in ruminants,' *Livestock Production Science 3,* 103.

MOE, P. W. and TYRRELL, H. F. (1973), 'The rationale of various energy systems for ruminants', *Journal of Animal Science 37,* 183.

ØRSKOV, E. R. and GRUBB, D. A. (1977), 'The effect of abomasal glucose or casein infusion on milk yield and milk composition in early lactation and negative energy balance', *Proceedings of the Nutrition Society 36,* 56A.

ROY, J. H. B., BALCH, C. C., MILLER, E. L., ØRSKOV, E. R. and SMITH, R. H. (1977), 'Calculation of the N-requirement for ruminants from nitrogen metabolism studies', *Protein Metabolism and Nutrition,* EAAP Publication No. 22, pp. 126–9.

Chapter 3

FEED INTAKE

The level of feed intake dictates the whole rate of metabolism in the body. Only in books and scientific experiments does the feeder set rigid feed allowances. In practice the feeder aims to satisfy his cattle's appetite and is soon reminded by bellowing and restlessness if he fails.

THE IMPORTANCE in cattle feeding of feed intake and the factors influencing it, are being increasingly realised. Gone are the days when books on livestock feeding dismissed the intake factor in a few sentences. Much further study is however required to enhance our ability to manipulate cattle feeding systems.

Factors influencing feed intake can be categorised as animal factors and feed factors although both factors are sometimes inseparable in their action.

ANIMAL FACTORS

Genetic Factors

Genotype exerts a major controlling influence on the complex inter-relationship of feed intake, body weight and thermogenesis (Owen, 1990). Genetic factors include differences between breeds and crosses and also the within-breed differences. Although there are large and obvious differences between animals of different breeds and crosses the breed effect becomes much less when body weight is taken into account. So far there is insufficient detailed information for confidence about the differences in feed intake between breeds when body weight is accounted for. However, as expected, some breeds and crosses do eat more than others in relation to their size.

The extra intake of breeds like the American Holstein and Dutch Friesian seems to be largely accounted for by the difference in body weight as indicated in Table 3.1.

Table 3.1. Intake and production in early lactation for North American Holsteins and Dutch Friesians given concentrates according to yield at two levels

	Breed (Number)			
	Holstein (12)		*Friesian (12)*	
Concentrate level	Low	High	Low	High
Concentrate DM consumed (kg)	9.1	16.1	8.5	15.1
Roughage DM consumed (kg)	11.7	7.1	10.5	6.4
Total DM intake (kg)	20.8	23.2	19.0	21.5
Body weight (kg)	616	622	583	583
Total DM intake per 100 kg body weight (kg)	3.4	3.7	3.3	3.7
Daily milk yield (kg)	31.1	32.4	25.4	26.5
Milk fat (%)	3.4	3.2	3.5	3.4

Source: Oldenbroek, J. K. (1979), *Livestock Production Science* 6, 147–51.

Size and Weight

Although it is clear that feed intake is related to body weight there are so many other influences involved that it is difficult to discern the precise form of the relationship.

Over a very wide range of species from mice to elephants it can be seen that feed intake shows a curvilinear association with mature body weight as shown in Fig. 3.1. This shows that doubling the weight of the species leads to less than a doubling of feed intake. The curve can be described approximately by saying that feed intake increases linearly in relation to body weight raised to the power of 0.75, i.e. $W^{0.75}$ or metabolic size. However too much is often made of this curve since in practice the range of mature weights is too small for the curvature to be detected. Although the surface of the earth is curved, because the world is round, for most practical purposes the curvature can be ignored. Also the fact that feed intake bears a certain relationship to mature body weight does not imply that the same relationship holds true for variations in weight of cattle according to age, for example.

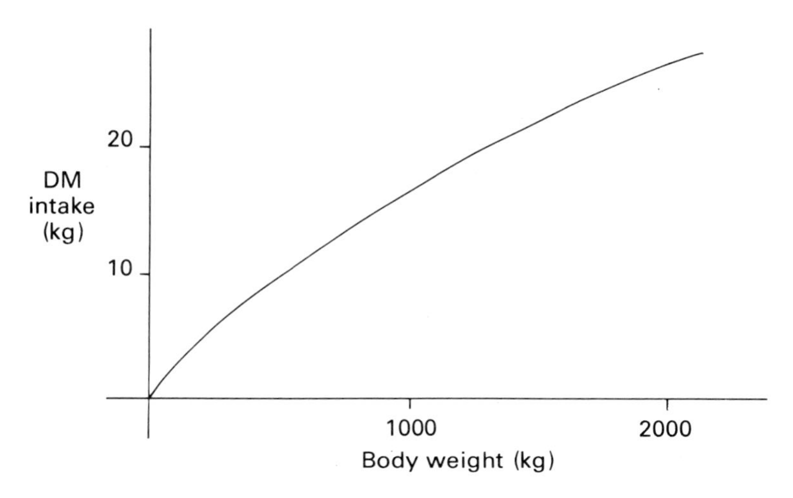

Figure 3.1. Relation of intake to bodyweight over a wide range (mice to elephants)
[*assuming $I = 0.09\ W^{0.75}$*]

Age

All measurable characteristics of cattle are associated with age although many other factors are superimposed on the basic age relationship. Fig. 3.2 illustrates the way feed intake changes according to age in steers kept on constant *ad libitum* regime. Feed intake changes with age in much the same way as body weight does. This is a sigmoid curve relationship where increases in intake with age are initially small because of the small size of the calf, increase to a maximum rate of increase at around puberty and then slacken off to reach a plateau or very small increase until senility sets in and intake begins to fall off.

Few cattle are allowed to demonstrate the whole of this range undisturbed. They may well be slaughtered at an early stage or, if a female, they will suffer a series of major disturbances associated with pregnancy and lactation.

It is interesting to note that the first-calf heifer not only shows a flatter lactation curve than the more mature cow but it also shows a similar type of feed intake pattern. Østergaard in Denmark studied Black and White cattle on *ad lib* silage

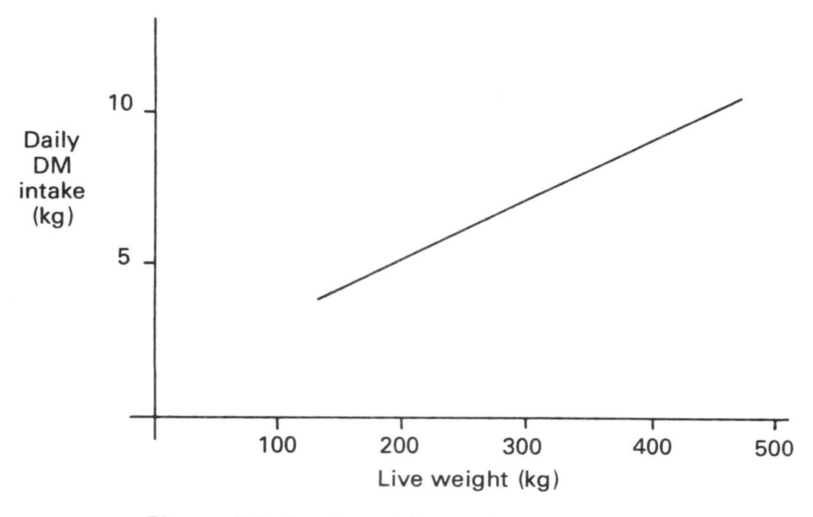

Figure 3.2. Intake with age in beef cattle
Source: Bowers, H. B. (1968) PhD thesis, University of Aberdeen

given in conjunction with concentrates. Overall in his studies cows averaged 538 kg and heifers 477 kg liveweight after calving a difference of 12.8 per cent. Total feed intake in the first 36 weeks of lactation was however only 3.8 per cent higher for cows, although in the first 24 weeks of lactation it was 5.1 per cent higher. This indicates that though heifers eat less in total they eat more than cows per unit of body weight and have a flatter feed intake curve with the peak intake occurring later in lactation than for the cow.

Such effects of age on the feed intake are seen at their greatest when comparing first-calf heifers with cows and the peak mature levels of intake are probably reached around the third or fourth lactation when the cow is 5–7 years of age.

Stage of Production

In the few cases where cows have been fed *ad libitum* on a standard diet for the whole of a production year, a clear pattern of intake is discernible as shown in Fig. 3.3.

This pattern, although present in ordinary feeding circumstances, tends to be modified by the changes in diet

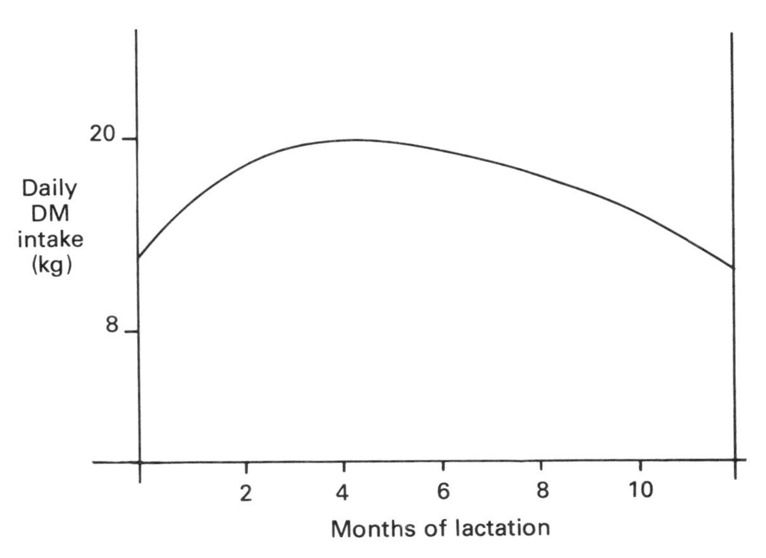

Figure 3.3. Pattern of DM intake of the cow over a lactation

that occur. A clear appreciation of this pattern of intake is important for the feeder since it lays down distinct limitations on how the diet is manipulated.

It shows firstly that intake is at its lowest ebb around calving and the day of calving usually marks the lowest intake of the year for a cow. Recovery from this low point is rapid under normal circumstances but the first two or three weeks are crucial since upsets at this time can easily precipitate metabolic disorders and any temporary bout of 'off-feed' can lead to more serious problems such as a misplaced abomasum or ketosis. A common fallacy is that since feed intake is low in the first few weeks of lactation, the energy intake can be easily increased by giving more concentrates. In fact, whilst feeding concentrates below a certain level can cause an unwanted reduction in energy intake, beyond this level increases in concentrates may have little influence on energy intake since total intake is even further depressed. It is salutary to remember that even on high-concentrate diets cows still show the low intake level and depressed energy intake characteristic of early lactation.

Peak intake usually occurs some weeks after the peak of lactation. This is a flattish peak maintained almost as a plateau until late lactation when the decline becomes more marked, particularly when the cow becomes dry.

The pattern of feed intake over the production cycle is modified, among other things, by the level of previous feeding. Garnsworthy and Topps (1982) demonstrated that cows which are fed on a high level, and consequently are in fat condition at calving, tend to eat less in early lactation and also tend to achieve a lower peak yield. Other things being equal it is always expected that cows in high condition will eat less than those in lower condition.

Level of Milk Yield

Most studies of the feed intake of cows have shown that the level of milk yield is a major factor influencing level of feed intake. Observations over a short term may not always reveal this close association because of the complicating effects of stage of lactation and other factors. When stage of lactation and other major factors are taken into account, however, level of milk yield usually ranks highest in importance as an influence on feed intake. Similarly whole-lactation studies also point to the same conclusion as illustrated in Table 3.2.

In general then, high-yielding cows tend to be big eaters, a point very well illustrated in a study, made by Albright of Purdue University, of the eating habits of the top world-record cows (Table 3.3). These cows seem to be fed on a reasonably normal diet in terms of forage:concentrate ratio but they stand out from normal cows as having a voracious capacity to consume food. Although it is not easy to decide which comes first—the urge to produce milk or the urge to eat—it seems more plausible to infer that the ability to eat has been stimulated by the high milk-producing tendency rather than vice versa.

To summarise the effect of milk yield on feed intake from the equations in Table 3.2 and others it appears that an increase of 1 kg in lactation yield elicits an average response of about 0.4 kg DM in annual feed DM intake.

Table 3.2. Equations relating DM intake of dairy cows to various measurable factors

1. $I = -4.14 + 0.43C + 0.015LW - 0.095\,WL + 4.04 \log WL + 0.208\,MY$
   ($R^2 = 71.9\%$ $n = 385$)

Source: Vadiveloo, J. and Holmes, W. (1979), *Journal of Agricultural Science 93,* 553–62.

2. $\ell n\, I = 0.52 - 0.00083DL + 0.148\, \ell n\, DL + 0.339\, \ell n\, MY + 0.0993\, FY + 0.000675LW + 0.018\, CF - 0.000557CF^2$
   ($R^2 = 74.1\%$ $n = 492$)

Source: Brown, C. A., Chandler, P. T. and Holter, J. B. (1977), *Journal of Dairy Science 60,* 1739–54.

Where I = Daily DM intake (kg)
C = Daily concentrate intake (kg)
LW = Liveweight (kg)
WL = Week of lactation
DL = Day of lactation
MY = Daily milk yield (kg)
FY = Daily butterfat yield (kg)
CF = Crude fibre as per cent of DM
ℓn = Natural logarithm

NOTE: Although these equations give reasonably good prediction overall they tend to under- or over-estimate intake at various stages of lactation.

Table 3.3. Feed consumption of World Record cow Ellen (1975) during her world record lactation (25,270 kg milk in 365 days on twice-a-day milking)

	Average daily consumption (kg DM)
Hay	24.5
Concentrates	26.3
Total feed	50.8
Concentrate: forage	42.48

Source: Albright, J. L. (1978), BOCM/Silcock Mimeo Report, 1978, 'The behaviour and management of high-yielding dairy cows'.

DIETARY FACTORS

A careful consideration of the animal factors influencing the feed intake of cattle indicates that each animal tends to strive to achieve a balance between intake and output, in the long term, that is consistent with the achievement of a genetically determined body size and composition in relation to age.

Every animal seems to behave as if it had a genetic blueprint which it can read and compare with an appreciation of its own body size and composition. If there is a discrepancy between the two in relation to age then several body mechanisms, including the chief one—feed intake level—respond in an attempt to correct the deviation. The extent to which this response can correct matters is much modified by the type of diet and is only fully achieved with an ideal diet of a fairly narrow range of specification. It is important for the feeder to know where this range lies since maximum production can only be achieved with such a diet. However it must also be remembered that such a diet is not always the most profitable diet to feed. Dietary factors include:

Energy Content

The intake of feed by cattle is markedly influenced by the digestible energy content of the diet although it is difficult to demonstrate this clearly in practice since the energy content of the diet cannot be changed without changing many of the other characteristics of the diet. Fig. 3.4 shows a diagrammatic illustration of the intake of cattle given diets varying in ME content. Two or even three phases can be seen in this association. Firstly at lower levels of ME, say 10 MJ/kg DM or below, intake is closely related to ME in a direct positive linear fashion. The higher the ME the higher the intake. On average the increase is something in the region of 5 kg DM intake for each increase of 1.0 unit of ME. The reason for this is well established and illustrates a distinct physical restriction on intake imposed by the limitation in the rate at which feed material is broken down and passes through the digestive tract. In most cattle diets that are given in the unground form, the main limiting factor is the rate at which the feed passes out of the rumen. As the energy content of the

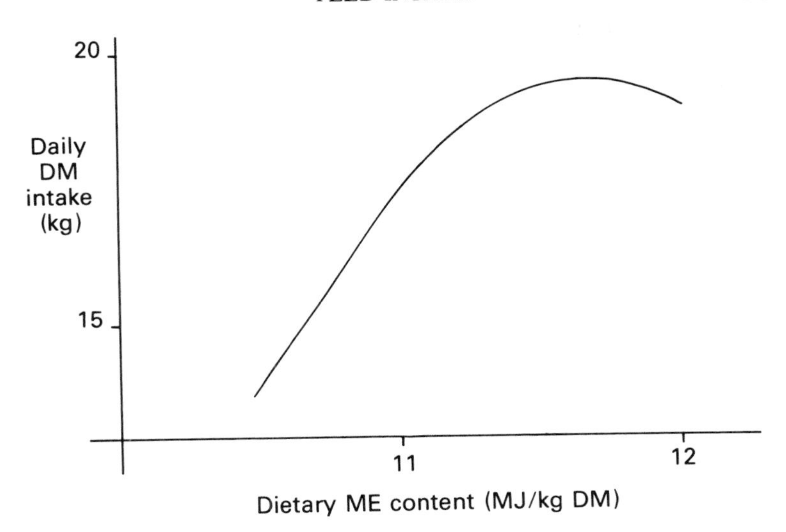

Figure 3.4. Intake of dairy cow in relation to dietary ME content
Source: Østergaard, V. (1979)

feed increases the diet is more quickly processed by chewing and microbial fermentation so that it can be absorbed or pass on to the rest of the digestive tract to make room for more.

At the higher levels of dietary energy the rate of passage of feed material out of the rumen ceases to be the limiting factor but the intake may still be physically limited, particularly on ground diets, by the rate of passage of material through the rest of the digestive tract. When even this limitation is removed the animal has then reached its maximum energy intake, which is metabolically determined by its genetic capacity for growth or milk yield. This is when the linear increase in intake with dietary ME level falls off and at the highest levels decreases occur. In between the two extremes is the phase where the response of intake to dietary ME level is weak and inconsistent and where there may often be little relation between ME level and intake.

This idealised relationship between intake and dietary energy level is not easy to appreciate in practice, particularly on the more concentrated diets, because of the marked influence of other factors that can overshadow this simple relationship.

Essential Nutrient Content of the Diet

All the remarks made about the association of feed intake and dietary energy level can be upset if the diet is of the wrong kind. For example, a deficiency of any of the essential nutrients—protein, vitamins and minerals—usually has an immediate depressing effect on intake. This might seem surprising at first sight since one might argue that if a nutrient is deficient in the diet the animal would be well advised to increase its overall intake in order to obtain more of the nutrient in question. However this is not the case since cattle need the essential nutrients in question to process the energy taken in via the food. The more they eat the more energy they absorb and have to metabolise and the more they require the essential nutrients to perform these tricky biochemical tasks. Energy absorbed by the body without the necessary complement of key processing ingredients is a serious embarrassment to the whole physiology of the animal. No wonder then that cattle are so made as to wisely reduce their total intake when essential nutrients are deficient. Nitrogen deficiency in the diet hits the animal in two ways; first by reducing the activity of the microbes so that their digestive functions are reduced; secondly by reducing the supply of amino acids for intestinal absorption. The first effect can reduce the rate of passage of feed and thus physically limit intake whilst the second effect has the direct metabolic effect on intake of a true amino acid deficiency.

Adding urea to a straw diet dramatically illustrates this point as shown in Table 3.4. The urea not only helps the microbes to break the straw down more quickly but indirectly

Table 3.4. Intake and digestibility of oat straw by steers as affected by the addition of urea supplements

Level of urea supplied (g/day)	*0*	*30*	*78*	*SE of mean*
DM intake (kg/day)	4.1	4.7	4.9	0.09
DM digestibility (%)	44	51	52	1.4
Crude fibre digestibility (%)	50	57	59	2.0

Source: Fishwick, G., Hemingway, R. G., Parkins, J. J. and Ritchie, N. S. (1973), *Animal Production 17*, 205–8.

increases the supply of amino acids to the intestine from the microbial bodies.

Similar effects have been demonstrated with the other essential nutrients—minerals and vitamins—but there is still a lack of knowledge about the levels of these nutrients that are necessary not only to prevent the major symptoms of deficiency but to ensure maximum intake and efficiency.

Excessive levels of certain constituents can also depress intake and there are few known chemical constituents of cattle diets that could not prove toxic in excess. Excessive levels of nutrients could affect intake even before overt toxic symptoms are evident, although at present the evidence for such effects is scarce because of the few studies that have been carried out.

Processing of the Diet and its Ingredients

Perhaps the major factor that can override some of the important dietary factors mentioned is processing. Processing can be applied in many forms; physical processing such as grinding and pelleting and chemical processing such as alkali treatment. Special processes such as those which protect the protein fraction have already been referred to and they can obviously affect feed intake through their effect on the protein value of the diet to cattle.

Grinding and pelleting forages. Since one of the major factors limiting the intake of many forages is the rate of passage out of the rumen it is not surprising to find that grinding the forage, so that particle size is much reduced, can markedly affect intake. With low-protein materials like cereal straws the effect is much greater if additional protein or non-protein nitrogen is also available as shown in Table 2.3. It is also greater the lesser the quality of the forage.

So great is the effect of grinding under appropriate circumstances that low-grade material like cereal straw, which is normally quite inadequate as a sole diet for cattle, becomes a feed capable of not only maintaining the animal but giving some liveweight gain.

Pelleting or cubing the diet is usually accomplished after grinding and adds only modestly to the transformation achieved by grinding. The pelleting process does involve a further physical breakdown of the material similar to the grinding process and it also condenses the material into dense packets that are easily consumed and avoid the dustiness that is associated with finely ground material.

The disadvantage of grinding and pelleting stems from the very phenomenon that gives rise to the greater intake—the faster rate of passage. This reduces the time available for ruminal fermentation and can significantly reduce the digestibility of the material. Thus some, but not all, of the value of the increased rate of intake is lost through a lowered digestibility.

Another feature of grinding and pelleting is the effect on the 'roughage' characteristics of the diet, which particularly concerns the dairy cow. Ground forage loses the physical properties of roughage and also changes the rumen fermentation pattern, such that it is less suited for lactation, although it may be beneficial for the growing beef beast.

Chemical treatment of forages has long been studied and practised to a limited extent—particularly the alkali treatment of cereal straws and similar materials. As shown in Table 3.5 such treatment increases the intake of feed by cattle and, unlike grinding, it also increases digestibility. Like grinding, the effect increases as the quality of the material decreases although the final quality of material that is initially low cannot be as high as better-quality initial material that shows lesser response to treatment.

Table 3.5. The effect on intake and digestibility of long and ground straw of treating with alkali

	Long straw		*Ground straw*	
	Untreated	*Treated with alkali (NaOH)*	*Untreated*	*Treated with alkali (NaOH)*
DM digestibility (%)	45	61	45	64
Daily DM intake (g/kg $W^{0.75}$)	27	48	36	54

Source: Carmona, J. F. and Greenhalgh, J. F. D. (1972), *Journal of Agricultural Science 78,* 477.

The alkali helps break down the tough cell wall constituents and enables them to be more easily broken down, as well as exposing the more digestible materials to the digestive process. The term chemical or alkali upgrading of feed materials is commonly used (Plate 5).

Whether the process is profitable depends on the cost of processing, the effect on feeding value and the type of cattle production system in which the upgraded material is used. Under present economic conditions in the U.K. (1983) the process is only viable in special circumstances.

Processing non-forage components of the diet. It has already been shown that processing is more effective the lower the digestibility and the larger the particle size of the material concerned. It is not surprising then that effects of processing concentrates on intake are not so apparent. Processing cereal grains for example has significant effects on digestibility but effects on intake have not been much reported (Table 3.6).

Plate 5. Ammonia treatment of straw on the farm (*Dairy Farmer*)

Table 3.6. The effect of processing barley grain by rolling (bruising) on the performance of beef cattle

	No Supplement	*Barley supplement (2.5 kg DM barley per day)* Whole	Rolled
Daily gain to slaughter (kg)	0.33	0.55	0.77
Carcass weight (kg)	215	217	237
Percentage composition of carcass			
Lean	55	54	53
Fat	30	32	34
Bone	15	14	13

Source: Broadbent, P. J. (1976), *Animal Production 23,* 165–72.

Mixing the whole diet in a machine like a mixer waggon has sometimes given rise to increased intake as compared to feeding the ingredients separately. This could be the result of some processing (akin to grinding but less severe) which occurs between the revolving augers of the machine or to the compression of the diet that can occur. It could also be partly an effect of easier access to the material than under separate feeding of the materials.

Special Intake Characteristics

Some mention needs to be made of the intake characteristics of some feed materials which are not directly related to the factors dealt with above. These special characteristics can be taken at present to cover an extensive area of mystery where unexpected intake levels are recorded for certain materials for no clear reason. Some of these feeds are variously described as being palatable or, conversely, as having negative intake characteristics.

Silage is one such material and several experiments have shown significantly lower intakes for material when it is conserved as silage compared with when it is made into hay as shown in Table 3.7. The ARC in its 1980 review came to the conclusion, however, that when a large number of trials were taken into account there was no need to assume lower

Table 3.7. Intake and performance of steers on the same grass conserved either as barn-dried hay or as silage (kg/day)

Approximate mean liveweight of cattle (kg)	*SILAGE* Intake of fresh material	*SILAGE* DM intake	*SILAGE* Daily gain	*HAY* Intake of fresh material	*HAY* DM intake	*HAY* Daily gain
250	18.1	4.8	0.12	5.8	5.0	0.69
300	23.6	6.3	0.54	7.3	6.3	0.59

Source: Forbes, T. J. and Irwin, J. H. D. (1968), *Journal of the British Grassland Society 23*, 299–305.

intake for cattle if the silage was well made; sheep on the other hand did show depressed levels of intake on silage.

A great deal more work needs to be done to confirm which materials have intake characteristics significantly different from expectation and to establish the reasons for this variance. It is important to realise that these differences can occur between natural forages given in the fresh state as shown in Table 3.8 which summarises the work on grasses at the WPBS Aberystwyth.

Table 3.8. Dry-matter intake and liveweight gain of 12-month-old and 24-month-old cattle offered single-variety grass feeds of different species at similar levels of digestibility

Grasses	*In vivo DOMD (%)*	*Daily DM intake (g/kg $W^{0.73}$)* 12 month old	*Daily DM intake (g/kg $W^{0.73}$)* 24 month old	*Daily DM intake (g/kg $W^{0.73}$)* Mean	*Liveweight gain (g/day)* 12 month old	*Liveweight gain (g/day)* 24 month old	*Liveweight gain (g/day)* Mean
S24	68.5	75.0	75.4	75.2	402	420	411
S22	68.5	70.1	86.4	78.3	344	598	471
S37	68.5	76.5	83.7	80.1	621	902	761
S51	68.7	61.6	66.1	63.9	−65	33	−16

Source: Miles, D. G., Walters, R. G. K. and Evans, E. M. (1969), *Animal Production 11*, 19–28.

INTERACTION BETWEEN ANIMAL AND DIETARY FACTORS

An important practical issue is how the animal and diet factors interact since they are both at work together in the farm feeding situation. At first it appeared plausible to expect

Table 3.9. Intake of fresh-cut grass by identical twin cows, one lactating the other dry

	Lactating cows				Dry cows		
Period Weeks from calving	*Daily FCM yield (kg)*	*Live-weight (kg)*	*DM digestibility (%)*	*DM intake (kg/day)*	*Live-weight (kg)*	*DM digestibility (%)*	*DM intake (kg/day)*
1–12	16.0	333	76	10.5 ± 1.06	370	76	8.0 ± 1.10
13–24	14.9	350	71	13.0 ± 1.09	421	72	8.6 ± 1.01
25–36	11.2	356	65	11.5 ± 0.99	446	65	7.0 ± 0.94

Source: Hutton, J. B. (1963), *Proceedings of the New Zealand Society of Animal Production*.

Table 3.10. Guide to dry-matter intake of various classes of cattle on medium diets

Dairy cows				
Yield per lactation FCM (kg)	*Month of lactation*	*Daily DM intake (kg) for cows of liveweight (kg)* 400	500	600
5000	1	10	12	13
	3	13	15	18
	10	11	13	15
7000	1	11	13	14
	3	14	17	19
	10	12	15	16
Beef cows	1	8	10	11
	3	11	13	15
	10	9	11	13

Growing/finishing beef cattle			
Diet	*Daily DM intake (kg) for cattle of liveweight (kg)* 200	350	500
Medium silage	4.1	6.3	8.2
Silage & concentrates (60: 40 DM ratio)	5.7	8.2	10.1

that the apparently physical limitation imposed with certain diets would override and obliterate any animal factors. Table 3.9 shows some of the work done in New Zealand at the Ruakura Institute on intakes of dry and lactating cows.

A reasonable conclusion from these findings is that whilst

dietary factors can severely limit overall intake, they do not necessarily override animal factors, e.g. a milking cow will eat more than a dry cow on a range of diets varying in digestibility and forage content.

Guides to the intakes of cattle under differing circumstances are given in Table 3.10.

REFERENCES

BAILE, C. A. and FORBES, J. M. (1974), 'Control of feed intake and regulation of energy balance in ruminants', *Physiological Reviews 54*, 160–214.

BALCH, C. C. and CAMPLING, R. C. (1962), 'Regulation of voluntary food intake in ruminants', *Nutrition Abstracts and Reviews 32*, 669–86.

BINES, J. A. (1979), 'Voluntary food intake', *Feeding Strategy for the High-yielding Dairy Cow*, pp. 23–48, Granada Publishing Ltd.

BROWN, C. A., CHANDLER, P. T. and HOLTER, J. B. (1977), 'Development of predictive equations for milk yield and dry matter intake in lactating cows', *Journal of Dairy Science 60*, 1739–54.

CONRAD, H. R. PRATT, A. D. and HIBBS, J. W. (1964), 'Regulation of feed intake in dairy cows. Changes in importance of physical and physiological factors with increasing digestibility', *Journal of Dairy Science 47*, 54–62.

CURRAN, M. K. and HOLMES, W. (1970), 'Prediction of the voluntary intake of food by dairy cows: 2. Lactating grazing cows', *Animal Production 12*, 213–24.

EL-SHOBOKSHY, A. S., JONES, D. I. H., MARAI, I. F. M., OWEN, J. B. and PHILLIPS, C. J. C. (1989), 'New techniques in feed processing for cattle', *New Techniques in Cattle Production*, edited by C. J. C. Phillips, Butterworths, London.

FORBES, J. M. (1986), *The voluntary food intake of farm animals*, Butterworths, London.

GARNSWORTHY, P. C. and TOPPS, J. H. (1982), 'The effect of body condition of dairy cows at calving on their food intake and performance when given complete diets', *Animal Production 35*, 113–20.

JOURNET, M. and REDMOND, B. (1976), 'Physiological factors affecting the voluntary intake of feed by cows', *Livestock Production Science 3*, 129–46.

OLDENBROCK, J. K. (1979), 'Differences in the intake of roughage between cows of three breeds fed two levels of concentrates according to milk yield', *Livestock Production Science 6*, 147–51.

OWEN, J. B. (1979), 'Feed intake' in *Complete Diets for Cattle and Sheep*, Farming Press Ltd.

OWEN, J. B. (1990), 'Weight control and appetite: nature over nurture', *Animal Breeding Abstracts 57*, 583–91.

OWEN, J. B., MILLER, E. L. and BRIDGE, P. S. (1971), 'Complete diets given *ad libitum* to dairy cows: the effect of straw content and of cubing the diet', *Journal of Agricultural Science 77*, 195–202.

Chapter 4

ASSESSMENT OF FEEDS AND OPTIMAL DIETS

Successful manufacture of any product entails marrying the detailed specification with the raw materials available. The process of diet formulation and of rationing cattle is also two-fold. It involves the determination of the optimal composition of a diet for the particular circumstance and the assessment of available feed materials in the same terms.

THE OLDER concept of an animal's requirements is becoming outmoded. The question of what the animal may 'require' at any particular time may be irrelevant in the context of the question of what type of diet and what level of feeding give the most profitable long-term conversion of inputs into product. We know full well, for example, that the dairy cow's 'requirements' in early lactation cannot always be met and we also know that in most feeding situations the attempt to match the diet closely to short-term, rapid changes in 'requirements' is both impractical and unnecessary.

The assessment of the optimal diet for various cattle-feeding circumstances is made difficult because of the many attributes of the diet that are known to affect the efficiency of the production process. Furthermore these attributes may interact together in a complex way that makes the prediction of what happens when varying mixtures of ingredients are put together a difficult task.

The main attributes of importance in a cattle diet, as far as present knowledge is concerned, are as follows:

1. Energy value.
2. Protein value, i.e. ability to supply amino acids.

3. Mineral content.
4. Vitamins and other essential micronutrient content.
5. Intake characteristics.
6. Physical characteristics of importance in normal body functioning.
7. Health-promoting properties.

All these attributes are important when viewed singly and it is also clear that many, if not all of them, interact together so that they must be considered together.

Energy

The useful energy as far as cattle are concerned is that which eventually finishes up as milk, as body tissue and as the fuel energy required to keep the body functioning and moving. Included in the latter fraction is the energy required to sustain the massive turnover of material in the body. Even when in a stable mature state, the cells of the body are continually ageing and dying off and being replaced by others in a process which keeps the body perpetually young.

This useful energy has been traditionally termed *net energy*, being the net result of the initial *gross energy* which was present in the food material when eaten, after accounting for a number of losses that occur in the process of conversion of gross energy into useful energy. The gross energy (GE) of the food is the total energy that is reasonably releasable in the material, i.e. the combustible energy that can be released when it is burnt. There is potentially of course much more energy in the food, such as the nuclear energy if it were ever released. Energy content can be directly estimated by using a bomb calorimeter in which the food is ignited, completely burnt to ashes and the heat given off measured. The gross energy of various food materials is shown in Table 4.1.

No one would accept this as a useful value of the energy for feeding cattle; for example the GE of straw, sawdust and barley certainly do not indicate their value as cattle feed.

The main categories of loss that have to be accounted for in

Table 4.1. Gross energy of selected feed materials

	Gross energy (MJ/kg DM)
Spring barley straw	18.0
Maize straw	18.1
Winter wheat straw	17.7
Potatoes	17.6
Swedes	17.7
Pasture grass	18.6
Lucerne (alfalfa) early flower	17.6
Heather	19.7
Brushwood	18.8
High-quality grass silage	18.2
Maize silage	18.8
Grass hay	17.7
Dried grass, leafy	18.0
Barley grain	18.3
Maize	19.0
Wheat	18.4
Extracted soya bean meal	19.5
White fishmeal	17.8
Sawdust	16.3–18.4

assessing the net energy (NE) from the GE are those involved in:

- undigested residue—faeces, dung.
- methane gas—released during fermentation in the rumen and evacuated through belching.
- urine losses—a small fraction of the urine—urea—contains combustible energy.
- heat losses—the remainder of the gross energy that does not appear in milk and body tissue.

The heat loss requires some further explanation. It represents two elements. One is the energy used for the maintenance of body function and depending on how wide the term 'maintenance' is used, can include the energy used for moving around. All this energy is released as heat since energy is not destroyed when it is used but converted into another form.

The other part of the heat energy represents the energy required in the process of converting energy into milk, into body tissue or into the process of body maintenance.

We therefore have several steps intervening between GE and NE and they can be categorised as follows:

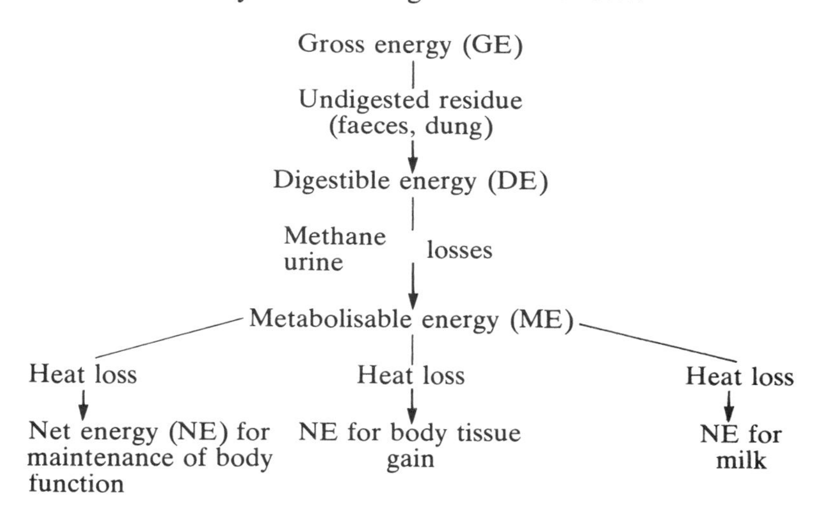

This is a slightly simplified representation although it is the main basis required for an understanding of cattle feeding. For example, cattle faeces not only contain undigested feed material but also a small fraction (the metabolic fraction) which consists of material such as undigested rumen microbes and the remains of secretions from the digestive tract such as digestive juices.

It is also common to subdivide the energy going into maintenance into that required by the cattle beast when resting (resting metabolism) and that required for activity.

This framework of the steps in energy use has been known since the turn of the century and has been the basis of many systems, at various times and in various countries, for the energy evaluation of foods and diets.

Because net energy is the final, useful fraction of the food, many early systems concentrated on using net energy units for expressing the energy value of feeds. Kellner's Starch Equivalent and Armsby's Thermal Unit are historic and well known. The Starch Equivalent (SE) System has only recently (in the 1970s) been replaced in the United Kingdom by the Metabolisable Energy (ME) System. Other countries, notably the Scandinavian countries, still use an NE system called the Fodder Unit System. The East Germans, particularly the Rostock scientists, have adapted and developed the NE system into a well-based practical system.

Although a system expressing feed value on the NE basis has an obvious practical relevance it was soon realised that it was not easy to operate in practice. In particular it was recognised that the heat losses in converting ME to NE were difficult to accommodate in a simple system. These vary markedly according to which of the three main end-products the ME is converted into (milk, body tissue, body maintenance), according to level of intake and the digestibility of the diet amongst other things.

The Americans realised this at an early stage and Morrison was chiefly responsible for concentrating much of the feed evaluation in the United States on the Digestible Energy (DE) fraction expressed as the Total Digestible Nutrients (TDN) value of the material. Using this method the main, easily assessed, loss in a feed was accounted for and it proved a simple, effective system for farmer use.

More recently, under the influence of Blaxter, the ME system has become widespread. This is similar to the TDN System but takes into account the methane and urine losses to provide a somewhat more accurate picture of the energy of a feed that is available for the animal.

Used on their own, as a simple comparative measure of the energy value of a wide range of feed materials, there is little to choose between the various units of energy in use. Most of the values in use for each energy system are derived from similar, mainly digestibility data, by rough conversion factors. None of them provide, in the one figure, a satisfactory evaluation when vastly different types of material, e.g. forage and cereal grains, are compared and when the type of cattle feeding system is not closely specified. However they are all much more useful guides when comparing different materials within a reasonably narrow category in relation to a specific feeding system. Furthermore, for such use there is no significant improvement in predictive value to be gained in using any value more sophisticated than the simplest—the DM digestibility value. There is, though, a strong argument for uniformity within a country or wider grouping so that only one rather than a multiplicity of units are used. In the United Kingdom, for instance, there is no point in using both D value (which is the digestible organic matter content of a feed expressed as a percentage of the total DM) *and* ME value for expressing the energy value of forages, as is the present practice. In the U.K. since the ME system has been widely adopted it seems reasonable to use the ME value in preference to others.

Table 4.2 summarises the ME value of a number of feed materials commonly used in cattle feeding. The materials have been arranged alphabetically within several broad categories. Within these categories materials can reasonably be compared in terms of energy value with others within the same category for the same purposes. Simple comparisons between materials in different categories are not however valid. This may not be such a drawback in practice as might be thought since many cattle-feeding situations lend themselves to a two-stage process where the first stage may involve a decision on the ratio of the two main categories of foodstuff—forage and concentrate—and the second stage then involves the decision on the optimum choice of ingredients within each of the two main categories.

Protein

Methods of expressing the protein value of feed materials also have a long history, but unlike the energy units which, in spite of proliferation, mainly estimate the same attribute with very similar accuracy, protein evaluation has undergone major and significant change. The basic first measure of protein value (somewhat akin to the gross energy value for energy) is the *crude protein* value which indicates the total potential protein value of the material. It would be more sensible and accurate to call it simply the ‘nitrogen content’ of the material but it has become conventional to multiply the nitrogen (N)

Table 4.2. Typical nutritive values of some common feedingstuffs used in cattle diets

Food: Category name (alphabetical order within category)	*Dry matter content g/kg*	*ME MJ/kg DM*	*Crude protein g/kg DM*	*RDP* g/kg DM*	*UDP* g/kg DM*	*Ether extract g/kg DM*	*Crude fibre g/kg DM*	*Calcium g/kg DM*	*Phos-phorus g/kg DM*	*Mag-nesium g/kg DM*	*Sodium g/kg DM*
Silage											
Barley (whole crop)	300	9.6	95	76	19	22	250	3.6	2.8	1.1	0.7
Brewers' grains	280	11.0	204	143	61	64	189	4.6	6.1	1.8	0.4
Clover (Red)	220	8.8	205	123	82	55	300	—	—	—	—
Distillers' grains	300	11.8	320	224	96	116	136	—	—	—	—
Grass (medium quality)	270	10.0	160	135	25	40	340	7.5	3.5	1.5	1.0
Lucerne (alfalfa)	250	8.5	168	101	67	84	296		—	—	—
Maize (corn)	210	10.8	110	66	44	67	233	[illegible]	2.1	1.8	—
Oats (whole crop)	240	8.0	79	63	16	33	358	—	—	—	—
Pea haulm	210	8.7	167	134	33	67	290		—	—	—
Sugar beet pulp	120	9.7	83	66	17	17	200	6.1	1.1	1.4	—
Hay											
Clover (Red)	850	8.9	161	129	32	35	287	—	—	—	—
Grass (good quality)	850	9.0	101	81	20	16	320	4.0	2.5	1.2	0.8
Lucern (half flower)	850	8.2	225	180	45	13	302	19.3	2.6	—	—
Millet	850	8.4	125	100	25	26	339	—	—	—	—
Artificially dried											
Grass (good quality)	900	10.8	161	97	64	28	217	8.0	3.4	—	—
Lucerne (early flower)	900	8.7	178	72	106	27	269	—	—	—	—
Straw											
Barley (spring)	860	7.3	38	30	8	21	394	4.2	0.8	—	—
Maize	850	7.3	59	47	12	18	461				
Oats (spring)	860	6.7	34	27	7	22	394	2.4	0.9	—	—
Rice (husks)	900	2.5	42	34	8	16	421	—	—	—	—
Soya bean	840	7.5	88	70	18	24	311	—	—	—	—
Wheat (winter)	860	5.7	24	19	5	15	426	—	—	—	—
Roots											
Fodder beet	230	13.7	48	38	10	4	48	1.6	1.9	1.0	2.0
Mangels	120	12.4	83	66	17	8	58	—	—	—	—
Potatoes	210	12.5	90	72	18	5	38	0.9	2.8	0.7	0.6
Sugar beet tops	160	9.9	125	100	25	31	100	8.6	2.8	—	—
Swedes	120	12.8	108	86	22	17	100	—	—	—	—
Turnips	90	11.2	122	98	24	22	111	4.8	3.4	1.1	5.1

Grains											
Barley	860	13.7	108	86	22	17	53	0.5	3.8	1.3	0.2
Maize	860	14.2	98	59	39	42	24	0.2	2.7	1.0	0.1
Millet	860	11.3	121	97	24	44	93	0.6	3.1	1.7	0.5
Oats	860	11.5	109	87	22	49	121	0.9	3.7	1.3	0.2
Rice (polished)	860	15.0	77	62	15	5	17	0.8	12.8	—	—
Rye	860	14.0	133	106	27	20	22	0.6	4.7	1.4	0.2
Sorghum	860	13.4	108	65	43	43	21	0.3	2.8	2.2	0.1
Wheat	860	14.0	124	99	25	19	26	0.3	4.0	1.2	0.1
Legume grains											
Beans (field)	860	12.8	290	232	58	15	85	1.9	6.7	1.3	0.2
Beans (Navy)	860	12.5	254	204	50	13	47	—	—	—	—
Peas	860	13.4	262	157	105	19	63	0.8	4.5	—	—
Byproducts											
Blood meal	900	13.2	942	754	188	9	0	0.4	2.2	0.3	9.7
Brewers' grains (dried)	900	10.3	204	122	82	71	169	3.2	7.8	—	—
Coconut cake	900	13.0	236	142	94	81	127	1.2	6.1	—	—
Cotton cake (decort)	900	12.3	457	274	183	89	87	3.2	14.7	—	—
Distillers' grains (dried)	900	12.1	301	181	120	126	110	3.1	3.3	—	—
Fishmeal (white)	900	11.1	701	421	280	40	0	79.3	43.7	2.2	16.1
Groundnut cake (decort)	900	12.9	504	404	100	67	72	1.6	6.3	2.4	0.8
Hominy feed	900	14.7	118	94	24	89	49	0.6	5.8	2.6	7.4
Linseed cake	900	13.4	332	199	133	107	102	4.0	8.2	6.1	1.6
Maize (flaked)	900	15.0	110	66	44	49	17	—	2.9	—	—
Maize germ meal (low fat)	900	13.2	112	90	22	36	34	2.1	6.7	—	—
Maize bran	900	12.5	96	77	19	47	132				
Maize gluten meal	900	14.2	394	315	79	52	23	0.4	1.4	—	—
Meat meal (low fat)	900	11.1	717	430	287	31	0	63.3	34.9	—	—
Meat and bone meal	900	7.9	527	211	316	44	0	114.4	60.0	2.2	—
Molasses (cane)	750	12.7	41	33	8	0	0	7.0	0.7	—	—
Molasses (beet)	750	12.9	47	38	9	0	0	3.5	0.3	—	—
Oat husks	900	4.9	21	17	4	11	351				
Palm kernel cake	900	12.8	216	173	43	68	150	2.3	5.3	—	—
Soya bean meal (extracted)	900	12.3	503	302	201	17	58	2.3	10.2	3.1	5.0
Sugar beet pulp (dried and molassed)	900	12.2	106	64	42	6	144	6.3	0.7	1.6	3.0
Tapioca flour	900	15.0	20	16	4	6	29	1.8	1.2	—	—
Wheat bran	880	10.1	170	136	34	45	114	1.6	8.4	5.0	1.3
Wheat middlings	880	11.9	176	141	35	41	86	[illegible]	9.1	3.9	0.5
Yeast (dried)	900	11.7	443	355	88	11	2	[illegible]	27.9	11.9	1.2

* RDP and UDP values provisional.

value by a factor of 6.25 and to call it crude protein. This is because, if all the nitrogen were in the form of protein, such a value would approximate closely to the protein content of the material since most common proteins contain about 16 per cent N, i.e. $\frac{1}{6.25}$. Although many cattle foodstuffs contain nitrogen compounds other than protein and proteins differ in their value (because of differing amino acid content) the nitrogen content or crude protein value of a material gives an indication of the maximum protein value of a material.

The next step in protein evaluation has been more difficult to accomplish than with energy. This is because the true digestibility of protein material is less easy to assess, since a larger proportion of nitrogen excreted in the faeces is of metabolic origin and derives from sources other than the undigested feed residue. Also, whilst digested energy has a positive, though somewhat variable effect on energy supply, digested protein may in certain circumstances simply reflect protein that has been degraded and proved of no value to the animal at all. Because of these considerations the digestible crude protein (DCP) value of a material is not very useful and the sooner it is discarded, in its present form, probably the better.

A significant step was the attempt to divide the crude protein (CP) value into two components—the rumen degradable protein (RDP) and the undegradable protein (UDP). The rumen micro-organisms, quite apart from their host cattle, have a requirement for protein, much of which fortunately can be synthesised from simpler nitrogen compounds notably ammonia (NH_3). The RDP component is the only part of the CP which can supply this need. Fortunately, provided the RDP supply is reasonably even and not derived from large meals of easily degradable material, then much of it can be used by the micro-organisms up to the limit of their capability for growth.

The UDP is the only fraction of the dietary CP that is potentially directly available to the host for 'normal' protein digestion. However some of this UDP is not only undegradable by the microbes but can withstand enzymic digestion in the abomasum (stomach) and intestines and may be excreted in the faeces. Also proteins vary in their amino acid content and UDP can have different value for the animal if it contains varying proportions of amino acids, particularly those more limiting, such as lysine and methionine.

Since the late seventies methods have been developed that can be used to estimate easily the rumen degradation properties of materials. At the University of Wales, Bangor, the method used has been the 'nylon bag' method, where samples of materials are placed in small nylon bags (like tea bags) and suspended through a fistula in the flank of cattle into the rumen. Several bags are suspended by threads at the same time and these samples are withdrawn systematically in order to assess how much material has been degraded after fixed time intervals. Table 4.3 summarises some results for common cattle feeding materials of various categories.

Whilst these methods are still in the process of development they are likely soon to enable at least a reasonably

Table 4.3. Protein degradability of hay and silage of a wide range of quality assessed on cattle at University College of North Wales, Bangor

Silage sample No.	*Estimated ME value (MJ/kg DM)*	*Degradability of protein (%) at 24 hours*	*Degradability of protein (%) when 90% of DM has been fermented*
1	10.5	92.7	94.5
2	9.1	84.1	88.3
3	9.7	80.6	86.2
4	7.4	92.2	92.7
5	8.9	81.6	82.7
6	8.8	85.8	87.7
7	6.5	85.9	85.9
8	5.6	67.8	70.1
Hay sample No.			
1	12.1	94.1	91.7
2	10.2	74.7	82.5
3	10.7	61.5	78.1
4	10.5	60.0	63.9
5	10.2	70.1	72.3
6	8.2	60.4	65.9

Source: Chamberlain, A. G. (1983), private communication.

rough categorisation of the CP of materials into UDP and RDP. Some of the main difficulties at present stem from the fact that in certain feeding circumstances, e.g. for high-yielding dairy cows, where the level of feed intake is high, differences in degradability may be less than in other circumstances, such as at maintenance level at which the estimates are commonly made.

At present the further elaboration of the value of the UDP according to its digestibility and biological value (amino acid content) is still awaiting further research work to provide useful practical values. Such work is likely to be of some practical significance in the feeding of dairy cows and young cattle.

The values for RDP and UDP that are shown in Table 4.2 are, for the reasons outlined, rather tentative but they seem more soundly based than the widely disseminated DCP values.

Minerals and Vitamins

Minerals and vitamins cover a wide range of chemical substances known to be needed by the body for its various functions. Some idea of these can be gleaned from looking at the detailed chemical composition of the body of cattle.

Some major elements such as calcium, phosphorus and magnesium are major components of body tissue, in particular bone. Others such as sulphur are components of other tissues, such as the protein of the muscles. Other elements are found in much smaller quantities but may still play a crucial role in some of the myriad of enzyme systems involved in body metabolism.

As far as these so-called minerals and vitamins are concerned their value in the feed material is expressed simply as their content in the material. Although there is evidence that the availability to cattle varies according to the type of feed material, the form in which the element is contained and the presence of other elements, no formal systematic way of incorporating this knowledge into practical feed evaluation has yet been found. Among the few exceptions are that ratios of calcium and phosphorus are commonly expressed.

References

Hansard, S. L. (1983), 'Microminerals for ruminant animals', *Nutrition Abstracts and Reviews 31,* 1–24.

Henderson, A. R., Evart, J. M. and Robertson, G. M. (1979), 'Studies on the aerobic stability of commercial silages', *J. Sci. Food Agric. 30,* 223–8.

Ministry of Agriculture, Fisheries and Food (1976), 'Nutrient allowances and composition of feeding stuffs for ruminants', *ADAS Advisory Paper No. 11.*

Owen, E. (1979), 'Processing of roughages' in *Recent Advances in Animal Nutrition—1978,* pp. 127–48, Eds., W. Haresign and D. Lewis, Butterworths, London.

Rowett Research Institute (1981), 'Feedingstuffs Evaluation Unit, Third Report', Department of Agriculture and Fisheries for Scotland.

Chapter 5

DIET FORMULATION

It is one thing to be able to assess feedingstuffs in relation to their feeding value but it requires a further step before such knowledge becomes useful in feeding practice. That step is to determine the specification of a diet, made up from one or several ingredients, that gives the best profit to the feeder.

Ideally what is required for diet formulation is:

(a) A list of available materials.
(b) Nutritive value of materials (cf Table 4.2).
(c) Function relating food intake to nutritive value of diet.
(d) Cost per unit of material and per unit of output.
(e) Input/output functions relating the output of product(s) to inputs.

Up to the present day it is step (e) that has caused the greatest difficulty for cattle feeding. One key feature in trying to reach the best solution is to consider the general form of possible input/output relationships as shown in Fig. 5.1 (a) and (b). Where the relation of output to input is as in 5.1 (a) it seems reasonable in many cases to aim for A as the ideal specification for the level of that nutrient in the diet. Up to A increasing the input gives a progressive response in output, but after A there is no further response. Point A can be regarded naturally as a 'requirement' for a specific nutrient and the most profitable diet will contain somewhere near that level. In practice, since many nutrients are being regarded simultaneously, the situation is less simple although the cost of providing some of the nutrients is small enough, relative to the total cost of the diet, that excess can be provided for safety.

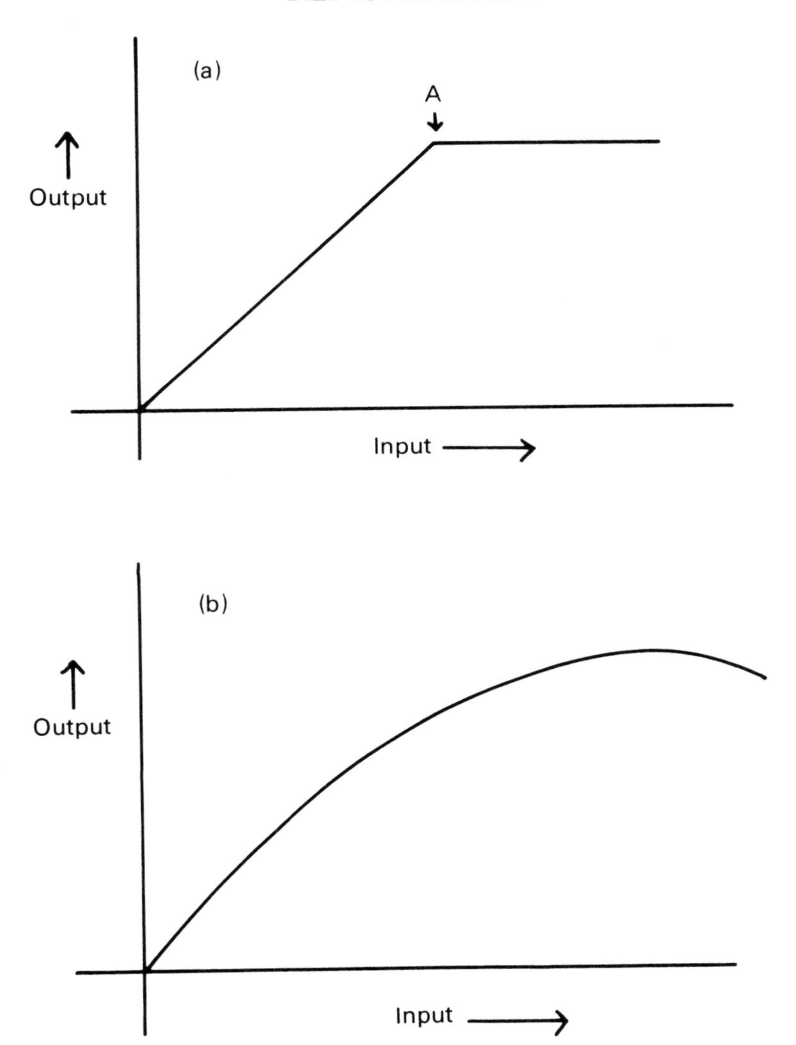

Figure 5.1. Forms of input/output

Unfortunately the response to the major components of the diet is not this simple; in fact it is usually of the form shown in Fig. 5.1 (b) where the response becomes curvilinear and tapers off gradually to an indeterminate maximum or to fall again. This has been known for a long time and a few lone voices such as Jawetz (1961), the Aberystwyth agricultural economist, have stressed the importance for dairy cow feeding.

In these circumstances there is no static specification for a diet based on so-called 'requirements' but an optimum value for nutrient level that depends on several factors, not least the cost per unit of feed materials.

When scientists first tackled the problem of feed rationing it seemed reasonable to attempt to assess what energy, for example, a cow or a steer needed every day, at a certain level of production and to meet those needs.

Taking dairy cows as an example, it is possible to assess the energy output in the form of milk and the energy needs of the cow for maintenance and activity. The latter have been determined using sophisticated metabolism chambers where heat losses of resting cows have been directly determined. Putting all these values together it is then possible to state what total sum of energy has to be provided daily to keep the cow in balance or, to give the system more flexibility, in some pre-determined slight positive or negative balance.

This approach, embodied in such publications as *Nutrient Requirements for Ruminants* (ARC, 1980) is a good first approach to the problem of formulation *but* there is grave danger in regarding it as anything more than an unsatisfactory first approximation. Its weakness is that it is entirely static and does not explore the farmer's prime question, 'What level is the most profitable for me to feed concentrates?' The farmer knows full well that, although the requirement for a litre of milk may be 0.4 kg of dairy cake, if he increases his concentrate allowance to his herd by 0.4 kg cake per cow he often gets nothing like such a return.

In this chapter an attempt is made to take the more positive dynamic approach to feed formulation using what scarce data are available. It is imperative however that the cattle industry ensures that resources are available to fill in the many gaps in this vital area of knowledge so that the approach can be developed and refined.

Formulation of Beef Cattle Diets

The simpler problem is taken first, since beef cattle are kept solely for meat production. At the present state of knowledge, the simplest procedure in formulating beef cattle diets, suitable for cattle from three months old upwards, is to deal with the diet as a two-fold entity—forage and concentrates. Although the line delineating the two components of the diet is not always so clear there is sufficient distinction to make it a useful practical division. The first step in formulating an optimum diet would then be to assess the optimum combination of forage and concentrates assuming forage and concentrates of average quality. Adjustments can then be made within each of the two broad categories to optimise the best combination of available materials to make up each category. Often, of course, the forage material is already pre-set and there is no choice to be made. If, after this process, it is obvious that the forage or the concentrates are not of the assumed average quality, adjustments need to be made to allow for this.

Thus to summarise we have a three-step process:

1. Establish optimum forage content assuming average value for both forage and concentrates.
2. Determine the least-cost combination of materials to supply the concentrate fraction (roughage also if choice is available).
3. Adjust forage content according to deviation from average quality.

Step 1. In establishing the optimum forage content it is convenient to divide the beef-finishing process into three stages assuming an end-product as a 600 kg live steer finishing at approximately 24 months of age.

Stage 1. 3–9 months 100–250 kg
Stage 2. 9–15 months 250–400 kg
Stage 3. 15–24 months 400–600 kg.

Fig. 5.2 summarises the results of studies on diets for beef feeding on the northern European diet of grass silage and

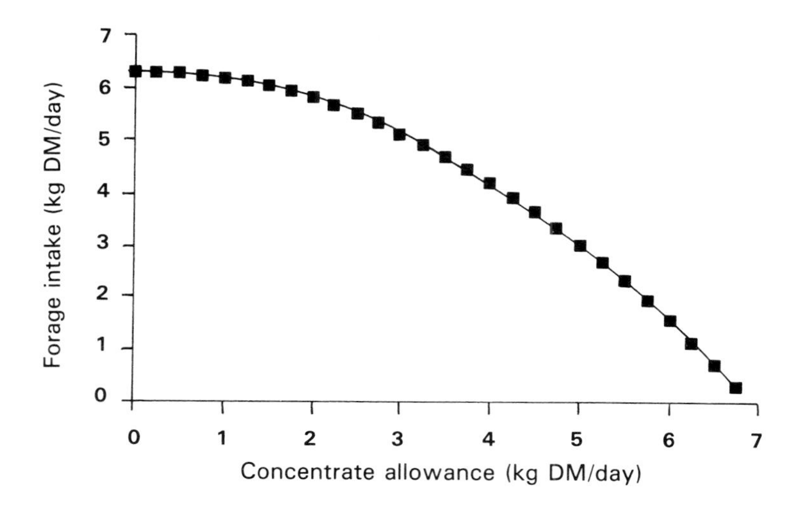

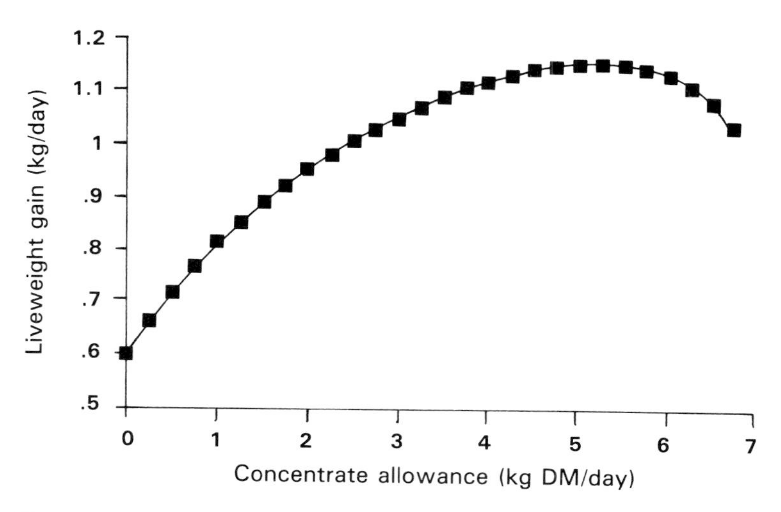

Figure 5.2. Forage intake and liveweight gain of 350 kg cattle in relation to concentrate intake (based on Table 5.1)

barley. These are also shown as equations in Table 5.1. These data show the daily liveweight gain and daily forage DM intake over a range of concentrate inputs. In Table 5.2 these data are incorporated, with several assumptions about the ratio of the price of forage DM to concentrate DM, and value of gain, to derive the optimum forage content of the diet.

The optimum forage content is taken as that which maximises the margin of value of gain over feed cost. Fortunately this optimum is not sensitive to the rate of gain on basal forage only. However at the lower end of the range of concentrate input the values in Table 5.2 may underestimate the real optimum since the effect of overhead costs may be reduced because of the higher rates of weight gain.

Step 2. In many circumstances there is no choice as far as forage is concerned although it might be possible to consider replacing some silage by grass or legume hay. If this were so then the decision could be based on a comparison of the silage and hay on the basis of the cost of supplying ME, e.g. if hay of 8.5 MJ/kg DM of ME is available

Table 5.1. Within-experiment relationships between DM intake and liveweight gain and various independent variables used to derive the data of Table 5.2

$$DMI = I_o + I_c - 0.131\ (\pm\ 0.0070)I_c^2 \quad (R^2 = 0.92)$$

where DMI = intake of DM
I_o = intake of silage when no concentrates are given
I_c = concentrate allowance

$$G = G_o + 0.017(\pm 0.0021)\ CR - 0.00013(\pm 0.000024)CR^2 \quad (R^2 = 0.90)$$

where G = daily liveweight gain
G_o = daily gain on silage only
CR = ratio of concentrate DM to total DMI

Source: Derived from data presented by Forbes, T. J. and Irwin, J. H. D. (1970), *Journal of British Grassland Society 25*, 96–103, and by Petchey, A. M. and Broadbent, P. J. (1980), *Animal Production 31*, 251–7.

at £70 per tonne of 85 per cent DM content then grass silage of 10.0 MJ/kg DM of ME of 25 per cent DM content would need to be valued in excess of £25 per tonne for the purchase of hay to be clearly justified.

This is because the cost per megajoule of ME in hay is 7000 pence divided by 1000 kg × 0.85 × 8.5 which is $\frac{7000}{7225}$ = 0.97 pence per megajoule.

If the ME in silage is set at the same value then the value per tonne of fresh silage is 1000 × 0.25 × 10.0 × 0.97 = £24.25 per tonne.

A modest replacement of silage by hay, say up to 20 per cent on a dry matter basis, could be accommodated with little basic change in assumed forage quality since the intake characteristics of hay are known to be above average in relation to its digestibility as compared with silage and there is also evidence that silage/hay mixtures tend to give performance somewhat better than expected from their performance when given singly.

In formulating the concentrate fraction of the diet there is usually more choice than with the forage. The necessity now is to look at the possible ingredients and to formulate a least-cost combination that will supply the necessary nutrients at the least cost per unit of ME.

For this it is necessary to specify the optimum diet as set out in Table 5.3. The data in this table are based on the best estimate of the optimum derived from the published results of experiments with cattle (ARC, 1980). These data can now be used in conjunction with those in Table 4.2 and the cost of ingredients to formulate a least-cost diet.

Some examples using fairly simple situations are worked out below to show how the procedure works (Table 5.4).

Although it is important to work out simple examples by hand to grasp the principle, it can be appreciated that when more materials are available, for instance, that the calculations can become time-consuming especially with regard to the iterative (trial and error) procedure.

Fortunately this procedure has been mechanised, using linear programming and computing techniques, so that diets can be produced by a simple question and answer process. Table 5.5 shows the output from a computer programme

Table 5.2a. Optimum daily concentrate allowance* (kg DM)

Relative cost of forage per kg DM	*Cost of concentrate per kg DM relative to the value of each kg liveweight gain* .07	*.08*	*.09*	*.10*	*.11*	*.12*	*.13*	*.14*	*.15*	*.16*	
.10	4.5	4.25	4.25	4	3.75	3.5	3.25	2.75	2.5	2.25	
.09	4.25	4.25	4	3.75	3.5	3.25	3	2.5	2.25	2	
.08	4.25	4	3.75	3.5	3.25	3	2.75	2.5	2.25	2	*Cattle of*
.07	4	3.75	3.5	3.25	3	2.75	2.5	2.25	2	1.75	*200 kg*
.06	3.75	3.5	3.25	3	2.75	2.5	2.25	2.25	2	1.75	*liveweight*
.05	3.5	3.5	3.25	3	2.75	2.5	2.25	2	1.75	1.75	
.04	3.5	3.25	3	2.75	2.5	2.25	2	2	1.75	1.5	
.10	6.25	6	6	5.75	5.75	5.5	5.25	5	4.75	3.75	
.09	6	6	5.75	5.75	5.5	5.25	5	4.5	3.75	2.25	
.08	5.75	5.75	5.5	5.5	5.25	5	4.5	3.75	2.75	2	*Cattle of*
.07	5.75	5.5	5.25	5	4.75	4.5	3.75	3	2.25	1.75	*350 kg*
.06	5.5	3.5	5	4.75	4.25	3.75	3	2.5	2	1.5	*liveweight*
.05	5.25	5	4.75	4.25	3.75	3.25	2.5	2	1.75	1.5	
.04	4.75	4.5	4.25	3.75	3.25	2.75	2.25	2	1.5	1.25	
.10	7.25	7.25	7.25	7.25	7	7	6.75	6.75	6.5	6.25	
.09	7.25	7.25	7	7	6.75	6.75	6.5	6.25	6	3.5	
.08	7	7	6.75	6.75	6.5	6.25	6	5.75	3.5	1.25	*Cattle of*
.07	6.75	6.75	6.5	6.25	6	5.5	4.5	4.5	1.5	1	*500 kg*
.06	6.75	6.5	6.25	6	5.75	5.25	4.25	2	1.25	1	*liveweight*
.05	6.25	6.25	6	5.5	5	4	2.5	1.75	1.25	1	
.04	6	5.75	5.5	4.75	4	2.75	2	1.5	1	.75	

* These can be adjusted up or down by up to 1 kg if silage of poor or excellent quality is used.

to carry out these procedures at the University of Wales, Bangor. This programme has been adjusted to ensure that the solution gives the best value for money in terms of supplying ME, having satisfied all the other specifications.

Step 3. Having now estimated the forage content, assuming average quality for both forage and concentrate, some

Table 5.2b. Silage intake at optimum concentrate allowance (kg DM/day)

Relative cost of forage per kg DM	*Cost of concentrate per kg DM relative to the value of each kg liveweight gain*										
	.07	*.08*	*.09*	*.10*	*.11*	*.12*	*.13*	*.14*	*.15*	*.16*	
.10	1.4	1.7	1.7	2.0	2.3	2.5	2.7	3.1	3.3	3.4	
.09	1.7	1.7	2.0	2.3	2.5	2.7	2.9	3.3	3.4	3.6	
.08	1.7	2.0	2.3	2.5	2.7	2.9	3.1	3.3	3.4	3.6	*Cattle of*
.07	2.0	2.3	2.5	2.7	2.9	3.1	3.3	3.4	3.6	3.7	*200 kg*
.06	2.3	2.5	2.7	2.9	3.1	3.3	3.4	3.4	3.6	3.7	*liveweight*
.05	2.5	2.5	2.7	2.9	3.1	3.3	3.4	3.6	3.7	3.7	
.04	2.5	2.7	2.9	3.1	3.3	3.4	3.6	3.6	3.7	3.8	
.10	1.2	1.6	1.6	2.0	2.0	2.3	2.7	3.0	3.3	4.5	
.09	1.6	1.6	2.0	2.0	2.3	2.7	3.0	3.6	4.5	5.6	
.08	2.0	2.0	2.3	2.3	2.7	3.0	3.6	4.5	5.3	5.8	*Cattle of*
.07	2.0	2.3	2.7	3.0	3.3	3.6	4.5	5.1	5.6	5.9	*350 kg*
.06	2.3	4.7	3.0	3.3	3.9	4.5	5.1	5.5	5.8	6.0	*liveweight*
.05	2.7	3.0	3.3	3.9	4.5	4.9	5.5	5.8	5.9	6.0	
.04	3.3	3.6	3.9	4.5	4.9	5.3	5.6	5.8	6.0	6.1	
.10	1.3	1.3	1.3	1.3	1.8	1.8	2.2	2.2	2.7	3.1	
.09	1.3	1.3	1.8	1.8	2.2	2.2	2.7	3.1	3.5	6.6	
.08	1.8	1.8	2.2	2.2	2.7	3.1	3.5	3.9	6.6	8.0	*Cattle of*
.07	2.2	2.2	2.7	3.1	3.5	4.2	5.5	5.5	7.9	8.1	*500 kg*
.06	2.2	2.7	3.1	3.5	3.9	4.6	5.8	7.7	8.0	8.1	*liveweight*
.05	3.1	3.1	3.5	4.2	4.9	6.1	7.4	7.8	8.0	8.1	
.04	3.5	3.9	4.2	5.2	6.1	7.2	7.7	7.9	8.1	8.1	

adjustment needs to be made for any departures from this assumption. At present this can only be done on an approximate basis, taking into account the effect on both DM intake and on expected gain, of deviations from average

Table 5.2c. Optimum forage content* (kg per 100 kg of total diet)

Relative cost of forage per kg DM	*Cost of concentrate per kg DM relative to the value of each kg liveweight gain*										
	.07	*.08*	*.09*	*.10*	*.11*	*.12*	*.13*	*.14*	*.15*	*.16*	
.10	24	29	29	33	39	42	46	53	57	60	
.09	29	29	33	38	43	46	49	57	60	64	*Cattle of*
.08	29	33	38	42	48	49	53	57	60	64	*200 kg*
.07	33	38	42	46	52	53	57	60	64	68	*liveweight*
.06	38	42	46	49	55	57	60	60	64	68	
.05	42	42	46	49	55	57	60	64	68	68	
.04	42	46	49	53	59	60	64	68	68	72	
.10	16	21	21	26	26	30	34	38	41	54	
.09	21	21	26	26	31	34	38	45	54	71	
.08	26	26	30	30	35	38	45	54	66	74	*Cattle of*
.07	26	30	34	38	43	45	54	63	71	77	*350 kg*
.06	30	57	38	41	51	54	63	69	74	80	*liveweight*
.05	34	38	41	48	58	60	69	74	77	80	
.04	41	45	48	54	64	66	71	74	80	83	
.10	15	15	15	15	20	20	25	25	29	33	
.09	15	15	20	20	25	25	29	33	37	65	
.08	20	20	25	25	30	33	37	40	65	86	*Cattle of*
.07	25	25	29	33	39	44	55	55	84	89	*500 kg*
.06	25	29	33	37	42	47	58	79	86	89	*liveweight*
.05	33	33	37	44	55	60	75	82	86	89	
.04	37	40	44	52	69	72	79	84	89	92	

* These can be adjusted up or down by up to 10 if silage of excellent or poor quality is used.

quality material. These are summarised in Table 5.2 and the examples in Step 2 can now be completed to give the final overall result. This result is about the simplest procedure that can be used at present for combining the economic and

Table 5.3. Specification of minimum nutrient levels in diets for growing cattle

Liveweight (kg)	Target daily gain (kg)	Guide daily DM intake (kg)	PROTEIN (g/kg DM) Crude protein	RDP	UDP	MINERALS (g/kg DM) [illegible]	P	Mg	Na	VITAMINS (µg/kg DM) A	D
200	0.5	4.0	100	90	10	5.2	2.7	1.5	0.7	1200	6.0
	1.0	6.0	130	105	25						
350	0.5	6.0	90	90	0	4.1	2.7	1.5	0.7	1200	6.0
	1.0	8.0	110	110	0						
500	0.5	8.0	90	90	0	3.5	2.7	1.5	0.7	1200	6.0
	1.0	10.0	100	100	[illegible]						

Source: Adaptation of ARC (1980), *The Nutrient Requirement of Ruminant Livestock*, C.A.B.

Table 5.4. Formulation of diet for a beef feeding unit

Assumptions:
Average body weight of cattle = 500 kg
Value of 1 kg liveweight gain = £1.10
Available feeds:
grass silage, barley, wheat, soya bean meal, fish meal, mineral mix.
Nutritive value of feeds:
As in Table 4.2
Cost of feed ingredients (£/tonne DM):
silage = 70 barley = 130 wheat = 145
soya = 170 fishmeal = 340 mineral mix* = 300

Procedure:
If it is assumed that concentrate DM cost per tonne is likely to be slightly higher than barley for this type of cattle (say £135) the cost of a kg of concentrate relative to that of a kg of liveweight gain is .135/1.1 = .123 and for silage DM it is .070/1.1 = .064.

Table 5.2 shows that for values of .12 and .06 for relative concentrate & relative forage DM respectively, the optimum daily concentrate allowance for 500 kg cattle is 5.25 kg DM which corresponds to a daily silage intake of 4.6 kg DM and an optimum forage content of 47 kg per 100 kg of total diet.

In this example barley is the cheapest source of ME and since cattle of this weight have relatively modest protein requirements, common sense suggests that the following diet, composed of grass silage, barley and a minimal level of mineral mix, should suffice:

	Optimal diet (Kg/tonne)	
	DM	*Fresh wt*
Grass silage	470	740
Barley	528	259
Mineral mix	2	1
	1000	1000

Computer formulation (Table 5.5) confirms this.

* Supplying (kg/tonne) calcium 180 magnesium 50 phosphorus 60 sodium 96

technical information into a single meaningful answer for the feeder, which is consistent with generally accepted feeding standards.

If the feeder prepares the concentrate mix for his beef unit separate from the forage then the two portions of the overall diet can easily be separated after the above formulation procedure. For instance, in the examples worked out above,

Table 5.5. University of Wales diet formulation

	Feeds and their inclusion restraints	
	Inclusion (kg)	Inclusion (kg)
Feed	*(min.)*	*(max.)*
Silage	470	470
Barley	0	1000
Wheat	0	1000
Soya	0	1000
Fish	0	1000
Minerals	0	100

	Dietary Specifications (kg)	
Requirement	*(min.)*	*(max.)*
Weight	1000	1000
Energy	10.5	13
Protein	100	180
RDP	100	180
UDP	0	50
Oil	10	80
C. fibre	50	400
Calcium	3.5	10
Phos	2.7	8
Mag	1.5	4
Sodium	0.7	5

	Rounded Optimal Diet	
Feed	*Weight (kg)*	*Price (£)*
Barley	530	68.9
Silage	470	32.9
Minerals	2	0.6
Total	1002	102.4

	Nutrient Composition (Rounded Diet)	
Energy (ME)	11.94	MJ/kg
C. Protein	132.2	g/kg
RDP	108.8	,,
UDP	23.4	,,
Oil	27.8	,,
C. Fibre	168.8	,,
Calcium	4.1	,,
Phos	3.4	,,
Mag	1.5	,,
Sodium	0.8	,,

Table 5.6. Formulation of a concentrate mix to be given with a forage base to beef cattle.

Assumptions: as in Table 5.4.

Ideally the concentrate formulation should change with the level of concentrate given as in the complete diet example in table 5.4. In practice a concentrate of fixed formulation has to be adopted where the concentrate is given as a separate entity.

	Optimal formulation (kg/tonne DM)	
	Total diet	*Concentrate*
Forage	470	—
Barley	528	996
Minerals	2	4
	1000	1000

the concentrate fractions in the diets expressed on their own are shown in Table 5.6.

These values are worked out by expressing the weight of each concentrate ingredient as a fraction of the total concentrate.

Formulation of Cow Diets

The procedure for formulating diets for beef and dairy cows is similar in principle to that outlined for growing beef animals. It consists of the same three-step procedure and it can be applied, according to the farmers' wishes, to any system of concentrate allocation (discussed fully in Chapter 7).

One point that needs making at the start and it may also apply to the beef-finishing systems already dealt with, is that it has recently been confirmed by Danish workers that both the protein and fat content of the diet also may show economically important curvilinear response functions in relation to cattle feeding. These will therefore need to be incorporated with energy in more complex response functions as suggested by Jawetz (1961).

For the time being, because of the intensive work going on on the protein nutrition of cattle, this aspect has not been incorporated into the principles employed in this chapter.

The same three steps in formulating procedure can be outlined for dairy cows and a short consideration of particular circumstances for beef cows is included.

Step 1. In establishing the optimum forage content for feeding housed dairy cattle, some account must be taken of the systems employed. These vary from cattle units where cows are not allowed to graze at all, mainly in semi-arid areas of the USA and in Israel for example, to those where cows calve just before the grazing season and spend most of their lactation on grass. In between these extremes are units where cows calve in the autumn at the end of the main grazing season and therefore may spend more than half the lactation on winter feeding, before being turned out to grass again in the spring. Useful practical information of value in determining optimum forage content has been published, particularly by Østergaard in Denmark and Gordon in Northern Ireland.

For the no-grazing and autumn-calving situations the detailed results of Østergaard and his colleagues are particularly valuable (Østergaard, 1979) and are consistent with Gordon's results with autumn-calving cows (Gordon, 1982).

Østergaard and Thysen (1982) have set out a simple response function to optimise dietary forage content for cattle given a medium-quality forage diet of grass silage (10.2 ME MJ/kg DM).

Optimum concentrate allowance =

$$\frac{P_C - 2.48\,P_M}{0.068P_S - 0.236P_M}$$

where P_C = price of concentrates per kg DM
P_S = price of silage per kg DM
P_M = price of 4% FCM per kg.

This combined with an equation to estimate intake of similar quality silage:

$$I = 13.4 - 0.034I_C^2$$

have been incorporated into Table 5.7 which sets out the optimum daily concentrate intake, silage intake and forage content for a number of assumptions about the prices of forage and concentrates relative to that of fat corrected milk (F.C.M.).

Table 5.7. Economic optima for dairy cow diets at various ratios of milk price to forage and concentrate costs

Cost of silage DM relative to FCM	*Assuming price per kg FCM = 100* *Cost of concentrate DM relative to FCM*							
	40	*50*	*60*	*70*	*80*	*90*	*100*	
70	11.0	10.5	10.0	9.4	8.9	8.4	7.9	*Optimum concentrate allowance (kg/DM/day)**
60	10.7	10.1	9.6	9.1	8.6	8.1	7.6	
50	10.3	9.8	9.3	8.8	8.3	7.8	7.3	
40	10.0	9.5	9.0	8.5	8.0	7.6	7.1	
30	9.6	9.2	8.7	8.3	7.8	7.3	6.9	
20	9.4	8.9	8.5	8.0	7.6	7.1	6.7	
70	9.3	9.6	10.0	10.4	10.7	11.0	11.3	*Silage intake at optimum (kg DM/day)*
60	9.5	9.9	10.2	10.6	10.9	11.2	11.4	
50	9.8	10.1	10.5	10.8	11.0	11.3	11.6	
40	10.0	10.3	10.6	10.9	11.2	11.5	11.7	
30	10.2	10.5	10.8	11.1	11.3	11.6	11.8	
20	10.4	10.7	11.0	11.2	11.5	11.7	11.9	
70	46	48	50	52	55	57	59	*Forage content (kg/100 kg of total)**
60	47	49	52	54	56	58	60	
50	49	51	53	55	57	59	61	
40	50	52	54	56	58	60	62	
30	51	53	55	57	59	61	63	
20	53	55	56	58	60	62	64	

* Values to be adjusted up or down by up to 2 kg (concentrate allowance) or by up to 10 (forage content) if silage quality is either low or excellent compared with the average value assumed (10 ME).

Step 2. The process of formulating the least-cost combination of ingredients for the concentrate fraction now follows the same procedure as for beef cattle. The specifications of the diets appropriate for dairy cows are set out in Table 5.8.

Table 5.8. Specification of optimum nutrient levels in diets for cows

	Dairy cows in milk	*Suckling beef cows*	*Dry cows*
ME (MJ/kg DM)	*10.5–12.0*	*9.5–11.0*	*9.0–10.5*
Crude protein (g/kg DM)	140	130	100
UDP (g/kg DM)	40	30	10
RDP (g/kg DM)	100	100	90
Crude fibre (g/kg DM)	160–200	170–300	170–300
Ether extract (g/kg DM)	160–250	25–80	20–80
Calcium (g/kg DM)	5.0	5.0	3.5
Phosphorus (g/kg DM)	4.0	4.0	3.0
Magnesium (g/kg DM)	2.0	2.0	1.5
Sodium (g/kg DM)	1.5	1.5	1.0
Vitamin A (μg/kg DM)	1,500	3,500	2,000
Vitamin D (μg/kg DM)	15	15	20
Optimum forage content per 100 kg total diet		80–100	80–100

Table 5.9. Formulation of a diet for a dairy herd

Assumptions: Value of milk per kg FCM = £0.20
(i.e. £200 per tonne FCM)

Feeds available: As in Table 5.4
Cost of feeds: As in Table 5.4

On the prices assumed for the available feeds (Table 5.4) the cost of concentrates is likely to be about £140. Table 5.7 shows that if this cost is expressed in relation to FCM price as 100 i.e. (140/200) × 100 = 70 and the cost of silage DM is expressed in the same way (i.e. (70/200) × 100 = 35) optimum forage content of diet = 57 (actually between 56 and 57).

Hand calculation can be carried out fairly quickly by operators experienced enough to judge the likely outcome, otherwise computer formulation can be used (see Table 5.10).

	Optimal diet (kg/tonne)	
	DM	*Fresh wt.*
Silage	570	810
Barley	330	147
Soya	90	38
Minerals	10	5
	1000	1000

These are again based on a review of a large number of experiments as summarised by ARC (1980).

A major advantage of setting out these specifications in the form of proportions of the total diet, as opposed to specifying absolute daily amounts, is that much of the variation due to size of cow and level of production is thereby removed. Many of the nutrients are required in relation to the energy consumed so that large variations in nutrients needs due to the cow size and level of production virtually disappear on a dietary content basis.

The values from Table 5.8 can be used in conjunction with a list of available feedingstuffs and their prices, to carry out the same procedure as in the beef cattle section. The example opposite illustrates the method (Table 5.9).

Similar computerised linear programming techniques can be used in this case as well and the following computer output (Table 5.10) illustrates the form that the results are presented.

Step 3. Some adjustments can finally be made to the forage content, according to the quality of the forage and concentrates. At present these adjustments can only be made on a rather crude arbitrary basis but further knowledge and sophistication of the computer techniques will allow more accurate answers. A summary of the adjustments—based on effects on DM intake and yield—are shown in Table 5.7.

Again many farmers will wish to prepare and feed their concentrate separately from the forage (which may be available on a self-feed, *ad lib,* basis). This can be accomplished by expressing the concentrate ingredients on the basis of the total concentrate fraction and not of the whole diet as shown in Table 5.11.

Beef Cows

Since the milk of beef cows is not sold and they are normally a low-cost enterprise, a simpler approach can be used for formulating appropriate diets. This is done by adopting the levels of forage content in Table 5.8 and then completing Step 2 as for dairy cows.

Table 5.10. University of Wales diet formulation

Feeds and their inclusion restraints

Feed (min.)	*Inclusion (kg)*	*Feed (max.)*	*Inclusion (kg)*
Silage	570	Silage	570
Barley	0	Barley	1000
Wheat	0	Wheat	1000
Soya	0	Soya	1000
Fish	0	Fish	1000
Minerals	0	Minerals	100

Dietary Specifications

Requirement (kg) (min.)		*Requirement (kg) (max.)*	
Weight	1000	Weight	1000
Energy	10.5	Energy	12
Protein	140	Protein	190
RDP	100	RDP	180
UDP	40	UDP	80
Oil	30	Oil	80
C. Fibre	160	C. Fibre	250
Calcium	5	Calcium	10
Phos	4	Phos	8
Mag	2	Mag	5
Sodium	1.5	Sodium	5

Rounded Optimal Diet

Feed	*Weight (kg)*	*Price (£)*
Barley	330	42.9
Silage	570	39.9
Soya	90	15.3
Minerals	9	2.7
Total	1002	102.4

Nutrient Composition (Rounded Diet)

Energy (ME)	11.34 MJ/kg
C. Protein	172.3 g/kg
RDP	132.6 ,,
UDP	39.6 ,,
Oil	30 ,,
C. Fibre	193.9 ,,
Calcium	6.3 ,,
Phos	4.3 ,,
Mag	2 ,,
Sodium	2 ,,

Table 5.11. Formulation of a concentrate mix to be given with a forage base to dairy cows

Assumptions: as in Table 5.9

As in Table 5.6, dairy farmers may also practise separate feeding of concentrates and require a standard formulation for this purpose.

	Optimal formulation (kg/tonne DM)	
	Total diet	*Concentrate*
Forage	570	—
Barley	330	770
Soya	90	210
Minerals	10	20
	1000	1000

REFERENCES

ARC (1980), *The Nutrient Requirements of Ruminant Livestock*, Commonwealth Agricultural Bureaux.

BROSTER, W. H. BROSTER, V. J. and SMITH, T. (1969), 'Experiments on the nutrition of the dairy heifer. VIII Effect on milk production of level of feeding at two stages of the lactation', *Journal of Agricultural Science 72*, 229–45.

JAWETZ, M. B. (1961), 'Towards a theory of feeding dairy cows', *Journal of Agricultural Economics 14*, 280–307.

FIELD, A. C. (1981), 'Some thoughts on dietary requirements of macro-elements for ruminants', *Proceedings of the Nutrition Society 40*, 267–72.

GORDON, F. J. (1982), 'The effect of pattern of concentrate allocation on milk production from autumn-calving heifers', *Animal Production 34*, 55–61.

KAY, M., MASSIE, R. and MCDEARMID, A. (1971), 'Intensive beef production 12. Replacement of concentrates with chopped dried grass', *Animal Production 13*, 101–6.

ØSTERGAARD, V. (1979), *Strategies for concentrate feeding to attain optimum feeding level in high yielding dairy cows*, 482. Beretning far Stations Husdyrbrugs forsøg, Copenhagen, 138 pp.

ØSTERGAARD, V. and THYSEN, I. (1982), 'Strategy of feeding concentrates: How to optimise the ration to the dairy cow'. European Association of Animal Production, Leningrad.

PALMIQUIST, D. L. and CONRAD, H. R. (1978), 'High fat rations for dairy cows. Effects on feed intake, milk and fat production and plasma metabolites', *Journal of Dairy Science 61*, 890–901.

ROFFLER, R. E., SALTER, L. D., HARDIE, A. R. and TYLER, W. J. (1978), 'Influence of dietary protein concentration on milk production by dairy cattle in early lactation', *Journal of Dairy Science 61*, 1422–8.

STEEN, R. W. J. and MCILMOYLE (1982), 'Effect of animal size on the response in the performance of beef cattle to an improvement in silage quality', *Animal Production 34*, 301–8.

STRICKLAND, M. J. and BROSTER, W. H. (1981), 'The effect of different levels of nutrition at two stages of the lactation on milk production and live weight change in Friesian cows and heifers', *Journal of Agricultural Science 96*, 677–90.

VIRTANEN, A. I. ETTALA, T. and MÄKINEN, S. (1972), 'Milk production of cows on purified protein-free feed with urea and ammonium salts as the only nitrogen source and on non-purified feed with rising amounts of true protein', *Festskrift Til Knut Brierem*, pp. 249–76, Utgitt Av En Redaksjonskomite.

WHITTEMORE, C. T. (1980), *Lactation of the Dairy Cow*, Longman.

WRENN, T. R., BITMAN, J. WATERMAN, R. A., WEYANT, J. R., WOOD, D. L., STROZINSKI, L. L. and HOOVEN, N. W. JR. (1978), 'Feeding protected and unprotected tallow to lactating cows', *Journal of Dairy Science 61*, 49–58.

Chapter 6

CALF FEEDING

A real care for the welfare of the animals allied with common sense and attention to detail are still the main ingredients of successful calf rearing. Knowledge of the calf's digestive system and the type of feeds that are available for it can add considerably to the art of calf rearing developed through generations of experience and observation.

IT HAS already been shown that the calf in the early stages makes little use of its rumen, which later dominates its whole existence. In dealing with calf feeding we need to examine briefly three distinct stages:

(a) the first period of complete dependence on liquid feed;
(b) the period of transition from all-liquid to all-solid feeding;
(c) the early post-weaning period.

The length of these periods varies considerably according to the system of calf rearing adopted. For example, a potential stud bull calf may suckle a cow for up to 10–12 months, thus prolonging stages (a) and (b). At the other extreme the calf from the dairy herd is commonly weaned from liquid feed at 4–8 weeks of age.

Similar principles, however, apply to the process of feeding, whatever the system, although the skill of the formulator and the calf rearer are more severely tested on the early-weaning, artificial-rearing system.

MILK REPLACER FORMULATION

Although the calf must normally obtain its mother's milk

Plate 6. Calf with its mother for only a few hours after birth

(colostrum) in the first day of life, in an artificial rearing system it is essential to transfer the calf as soon as possible on to a milk replacer, on which it can achieve almost as high a performance as on natural cow's milk (Plate 6).

As already indicated the problem in milk replacer formulation stems from the rather restricted digestive enzyme system of the young calf in the first three to four weeks of life. Milk supplies the young calf with energy, protein (amino acids), vitamins and minerals; the milk replacer must do the same. Looking at the various fractions of whole cow's milk, consideration can be given to how they can best be replaced.

Protein

The main protein source in milk is casein and it represents

82 per cent of the total protein in milk. It not only has a high biological value for calf. i.e. contains a near optimum array of constituent amino acids, but it also has the unique role in milk clotting so essential for proper calf digestion and absorption. Experiments have shown that if calves are given feeds at very frequent intervals, such as twenty-four feeds per day the need for curd formation is less, since the calf's digestive system is reasonably evenly loaded over the twenty-four hours. Under bucket feeding conditions such a frequency is unattainable and even on self-feed, ad lib, sucking systems frequency of suckling is not such that the clotting mechanism can be dispensed with.

For normal circumstances, at the present state of knowledge, casein must therefore be the basis of the protein source in any milk replacer to be used below one month of age. This can normally be supplied from skimmed milk or butter milk. However several attempts have been made to replace part of the casein by other substitutes. Some likely candidates for such a replacement role like soya bean flour are not very suitable without treatment because they contain detrimental factors like certain enzyme inhibitors, e.g. trypsin inhibitors. However certain fish protein products have been used and experiments at Aberdeen have shown that under appropriate conditions replacement rates of up to 20–30 per cent of the casein can be achieved with little loss of performance. Such a replacement rate is better tolerated when the calf is given restricted feed levels. Such experimentation however has some difficulties and it is important not only to compare a 100 per cent casein treatment with say an 80 per cent casein/20 per cent fish mixture but to include the so called negative control of 80 per cent casein, in case the supply of protein to the calf in the experiment is slightly in excess, enough at least to account for the 20 per cent replacement.

As the calf gets older, notably three weeks appears to be something of a watershed, it becomes more robust and can utilise a wider range of substances. Since replacers are normally given for a period of five to six weeks, the most cost-effective replacer may be one that is slightly less than optimal in the first two to three weeks but is a cheap and effective replacer in the latter part of the liquid feeding period.

Carbohydrate

Lactose is the main carbohydrate in milk and is a quickly assimilable source of energy for the calf. Unfortunately the calf is sensitive to the level in the milk and levels above 40 per cent can cause scouring (diarrhoea). This is one of the reasons why straight skim milk and butter milk are unsuitable for rearing calves in the first three weeks of life although they can be tolerated after that age. Replacements for lactose are difficult in that the calf can only use glucose which is normally more expensive to provide than lactose. The calf in the early stages cannot cope well with more complex carbohydrates like starch. Normally therefore lactose, as supplied by milk by products and that in only limited amounts, is the only practicable carbohydrate source for milk replacers, suitable for the very young calf.

Fat

Butterfat is a key constituent of milk as food for the young calf because it is a concentrated source of energy and an important complement to the milk protein, since the calf's capacity for carbohydrate intake is so limited. Although butterfat is an excellent source of fat for the calf it is one of the milk constituents that can be replaced by a wide range of materials without reducing performance too much. Table 6.1 shows the digestibility of various fats to the young calf.

High-quality plant oils, such as those used in margarine manufacture, e.g. coconut oil, peanut oil and others, if properly treated, can provide fat in a replacer almost as efficiently as butter fat. Some of the animal fats, notably tallow, are not as well utilised but can, within limits, provide some of the fat in commercial milk replacers.

Perhaps more important in dealing with fat in calf milk replacers, is the way the fats are treated and incorporated into the replacer. To be fully utilised the fat must be broken down into very small globules (3–4 μ in diameter). This is achieved by the chemical process of emulsification using lecithin and/or the physical process of homogenisation. Because of

Table 6.1. Digestibility of fats from various sources in milk replacers for calves

	Mean apparent digestibility (%)
Butter fat	97
Coconut oil	95
Palm oil	93
Groundnut oil	93
Lard	92
Maize oil	88
Tallow	87

Source: Roy, J. H. B., Roche Information Service.

advances in homogenisation processes, plants are now quite capable of producing large amounts of high-fat powders, which mix well with water and are eminently suitable as calf milk replacers.

Minerals and Vitamins

There is normally no problem in incorporating these constituents since they form only a small part of the total but their efficient incorporation involves through mixing at a stage when heat processing is not going to inactivate some of the vitamins.

The Formulation Process

The next step is to put all the constituents together into a stable saleable powder for use in the calf-rearing unit. First of all the overall optimum composition of the replacer must be considered. Obviously a good first starting point in this respect is cow's milk itself, the composition of which is shown in Table 6.2.

Alongside this is shown the optimum composition of a milk replacer, recommended following considerable research into this subject over the years (Roy, 1980). One point of importance to remember in formulating milk replacers is that the level of feeding can be a factor. Milk replacers that may be suitable for restricted feeding may overload the digestive system and cause scouring when given ad libitum. Natural

Table 6.2. Approximate composition of colostrum, milk and high-quality milk replacer on a DM basis (g/kg)

	Colostrum	*Milk*	*Milk replacer*
Fat	273	313	200
Protein	571	242	249
Lactose	110	391	423
Minerals and other substances	46	54	128

milk and high-quality milk replacers on the other hand are suitable for all levels of feeding provided the calf is accustomed gradually to the higher levels.

The 'quality' of a milk replacer, particularly as far as the first three weeks of life are concerned, depends partly on the treatment of the milk in the manufacture of skim milk powder and of the milk replacer. Excessive heat treatment leads to denaturing of the whey proteins and results in a milk replacer that has poor clotting properties. Such a powder predisposes the calf to diarrhoea associated with the bacterium *E. coli* (white scours) and to reduced digestibility and utilisation of the replacer from a variety of causes. A useful indicator of the degree of denaturing that has occurred is the ratio of non casein N:total N. This is about 0.25 in fresh cow's milk and can reach as high as 0.22 in well-produced milk powders. Below a level of about 0.18 clotting properties are impaired and the calf's performance can be affected.

High-quality milk replacers, suitable for feeding in the first three weeks of life, are made up mainly of skim milk with added fat and a small proportion of other constituents; ideally not exceeding 20 per cent of the DM. These other ingredients could include whey and/or some of the better products based on soya bean, fish, meat or single-cell proteins grown on various substrates.

Such milk replacers can give good performance after the first few days of colostrum feeding without the necessity for incorporating antibiotics, which are not allowed in all countries.

Cheaper replacers, where more of the milk protein has been replaced by other products or more heat damage has occurred, must be fed with great care in the first month of life. The ill effects of lower quality milk replacers can be mitigated by the inclusion of antibiotics to control intestinal infections and by restricting the daily allowance.

Table 6.3 shows the composition of a typical 'high-quality' milk replacer.

Table 6.3. Composition of high-quality milk replacer suitable for young calves

	g/kg DM
Vegetable oil, e.g. coconut oil	200
Skim milk	650
Whey	100
Mineral vitamin addition*	50

* *This should contain per kg DM*

Magnesium	250mg
Iron	100mg
Manganese	40mg
Copper	10mg
Zinc	12mg
Cobalt	100μg
Iodine	120μg
Vitamin A	12,000–20,000iμ
Vitamin D	1,800–3,200iμ
Vitamin E	20mg
Vitamin B_{12}	30μg

Source: Roy, J. H. B. (1970), *The Calf,* Iliffe Books Ltd.

Acidified Milk Replacers

Many calf milk replacers are now treated with organic acids to produce high (pH 4.4) or medium (pH 5.6) acidified milk replacers. The process is claimed to increase the 'life' of the liquid replacer and to reduce the incidence of digestive upsets. Whilst the use of such replacers eases the feeding of calves under an ad libitum cold milk system, there is less evidence that they are superior when used for ordinary restricted feeding of warm milk.

Other Additives

In addition to the micronutrients listed in Table 6.3 several other substances have been advocated from time to time as profitable additions to the diet of the young calf.

Antibiotics can often lead to improved performance. They may help overcome the secondary bacterial infections that accompany digestive upsets in artificial rearing systems; they may also have other direct growth-promoting properties.

Several products are now available, known as probiotics, which are intended to encourage a healthy gut flora in calves and thus to reduce digestive upsets and improve performance. These products include preparations based on organisms such as *Lactobacillus acidophillus*, *Streptococcus faecium* and *Bacillus subtilis* and their fermentation products.

Although it has been demonstrated that products based on *B. subtilis* have resulted in reduced faecal coliform numbers, no convincing evidence is yet available that such probiotics are likely to prove profitable under normal calf rearing conditions (Shujaa, 1989).

Calf Starter Formulation

A calf starter is a solid diet that can be consumed by the calf before weaning and for the post-weaning period of transition to more normal growing cattle diets. Since the main aim of the diet is to sustain the calf in the traumatic period after liquid feeding has been stopped, it has to provide all the requirements of the calf via ruminal digestion.

The main features of such a diet are the proportion of high-quality roughage and the proportion of UDP. It also must supply the high levels of essential nutrients required by the growing calf as shown in Table 6.4.

The type of feed materials capable of providing such a diet are usually those that ensure a high intake level, i.e. are palatable. The way the diet is processed is also important, rolling being preferred to fine grinding.

Examples of materials that can provide the required type of diet are shown in Table 6.4. Fish meal and soya bean meal are examples of good protein sources because of their high UDP content. Cereals such as rolled barley and coarse ground maize are also suitable.

Table 6.4. Guide to optimum formulation of starter concentrates to be given in conjunction with hay suitable for young calves before and in the first weeks following weaning from a liquid diet

ME MJ/kg DM	*11.0–13.0*
Crude protein (g/kg DM) (min.)	160
UDP (g/kg DM)	60
RDP (g/kg DM)	100
Crude fibre (g/kg DM) (min.)	70
Calcium (g/kg DM)	7.0
Phosphorus (g/kg DM)	4.0
Magnesium (g/kg DM)	1.5
Sodium (g/kg DM)	0.8
Vitamin A (μg/kg DM)	1,000
Vitamin D (μg/kg DM)	5.0
Example starter diet to supply above needs	
	g/kg of mix
Maize (coarse ground)	300
Barley (rolled)	300
Dried grass meal	200
Fish meal	50
Soya bean meal	100
Vitamin/mineral mix	50

Diets that are formulated to provide sufficient UDP and the optimum roughage content usually supply sufficient RDP and energy for the calf.

The level of intake is crucial when monitoring the suitability of calf starter diets; this must be maximised to allow the calf to bridge the important gap from liquid to all-solid feeding.

Calf Feeding Systems

Natural Rearing

Calves born in specialised beef herds are normally suckled by the cow until the age of 6–9 months. Most of the calves are single suckled. Although it can be shown that double suckling (Plate 7) may make better use of resources, the problems of calf availability, the acceptance of strange calves by the cow and the lower performance of the individual calves (Table 6.5), usually militate against the widespread adoption of a double suckling system. Suckled calves are usually capable of

Plate 7. Double suckling in a commercial beef herd

Table 6.5. Effects of single- and double-suckling on calf performance

Liveweight (kg)	*Single-suckled calf*	*Double-suckled calf*
Birth/fostering	42.3	45.0
Weaning	216	177
Liveweight gain to weaning (kg/day)	0.91	0.68

Source: Adapted from Nicoll, G. B. (1982), *Animal Production 35,* 395–400.

a high level of growth and receive milk from the cow over a long period. Because they are consuming a good supply of milk protein direct into the abomasum, they are well nurtured in terms of protein. This enables them to maintain a high level of intake overall so that as they grow they increasingly supplement their diet by solid feed—often pasture grass. A useful ploy is to allow the calves access to creeps where they can consume a concentrate feed away from the cow. Creeps are used from an early stage for yarded cattle but on pasture they may only be employed in the period preceding weaning in the autumn when pasture quality and availability is falling in any case.

Artificial Rearing

The first days. Calves are normally removed from the cow within a few hours of birth. By this time the calf has sucked colostrum from the cow but the attachment of the cow to her calf is not so strong as to cause major withdrawal upsets.

Some cattlemen believe the upset is minimised if the calf can be removed before the cow licks the calf although this is often not a practical arrangement. Cows of some cattle breeds, usually those less developed by intense selection, are more upset by early calf removal than others and commonly show difficulties in milk let-down in the absence of the calf.

Colostrum. The cow's first milk—colostrum—is vital to calf survival since it supplies the calf with its first line of resistance against disease. This is provided by the absorption of immunoglobulins from the globulin fraction of the colostrum. These immune bodies are formed in response to the exposure of the cow to various organisms in its environment, particularly the *E. coli* types that it has to contend with. The calf is born with no ready-made resistance and its intestine is capable of absorbing these large immunoglobulin protein molecules in the first few hours after birth. This depends on the absence of enzymes that break down the protein at this stage and the permeability of the intestine to these molecules. After about twelve hours the immunoglobulins are no longer absorbed. The time-span is reduced if the calf has been given milk. It is important therefore to allow the calf as much colostrum as possible in the first six to eight hours after birth.

The colostrum also supplies other valuable necessities such as vitamins.

In many dairy herds enough colostrum is available to allow young calves colostrum for several days. Colostrum can be stored by keeping in a refrigerator (for a few days) or a freezer (several months) and it is also possible to prolong its shelf life by allowing it to become sour (through the production of lactic acid).

Ideally transition from cow's milk to milk replacer should be achieved gradually so that the calf is on milk replacer completely by the time it is seven to ten days old.

Several decisions have to be made with regard to milk feeding for the ensuing liquid feeding period:

- Level of feeding.
- Dilution rate of milk powder.
- Temperature of feeding.
- Frequency of feeding.
- Method of feeding—teat, bucket, machine etc.

Level of feeding. The level of feeding has to be a compromise between ad lib feeding, which will maximise growth rate of the calf in the milk feeding period but, because of a reduced level of solid food consumed, will increase feed cost and increase the weaning check. Several experiments indicate that because of the factors mentioned, the liveweight achieved at six months of age may be little different for a range of liquid feeding allowances.

For the rearing of both heifer replacements and of beef cattle the best policy seems to be that the liveweight gain target for the calf in the first three months should be at a medium level, around two-thirds of the maximum level. This entails the use of high-quality milk replacer given at 60–70 per cent of ad libitum level fed in conjunction with ad lib hay and calf starter (Table 6.6).

The detailed distribution of this level of milk replacer allowance over the six to seven weeks milk feeding period seems not to be critical but a common system is to build the calf up to its maximum daily allowance by about ten days of age and to retain this level until weaning.

Dilution rate. Experiments have shown that the calf has a highly developed appetite control system which enables it to adjust intake, over a reasonably wide range of dilution, so

Table 6.6. Guide to milk replacer allowances for artificially reared calves

Period (days)	*Type of feed*	*Level of feeding*	*Frequency of feeding per day*
0–4	Colostrum	to appetite (3.0 litres)	3x
5–10	Cow's milk } Milk replacer }	4.0 litres	2x
11–35/42	Milk replacer*	4.0 litres	2x

* At a concentration of 13 g milk powder DM per litre of liquid.

that energy intake remains the same. At the higher levels of dilution the calf is unable to adjust fully and it is recommended that concentration levels of 14 per cent or more are required for the maximum intake of low-fat milk replacers.

Temperature of the milk. The optimum temperature for feeding milk is in the 35 °–38 °C range, approximating the normal level experienced with natural suckling. Calves can be reared on cold milk, i.e. milk at the prevailing temperature, but performance is often affected. Observations have shown that under cold winter conditions calves show a variable response to feeding cold milk, even when high-quality liquid feed is available and liveweight gain and feed conversion efficiency can be affected.

However under less cold conditions and under appropriate management a cold milk self-feed system can be appropriate, particularly with acidified milk replacer. The feed cost on this system, like all ad lib systems, is higher than with moderately restricted milk feeding levels.

Frequency of Feeding

Calves normally choose their frequency of feeding on ad lib systems but on restricted artificial systems it is normal to give liquid feed twice a day. Many experimenters have shown that calves can be successfully reared on once-a-day feeding and that missing out a day's feeding on a normal twice-a-day

Plate 8. Calves teat-fed on cold acidified milk replacer

regime results only in quite small depression in performance.

Once-a-day feeding can be successfully adopted after the first week, provided the calves are on a restricted feeding system using good-quality milk replacer. It would also be advisable for this system to adopt a milk powder concentration of 15 per cent to facilitate consumption.

The system would not be advised for systems like veal calf production where high levels of intake are essential.

Method of Feeding

Many methods of feeding can be adopted in practice but the usual choice is between bucket feeding and teat feeding as shown in Plate 8. Teat feeding reduces the rate of milk consumption and can thus increase the rate of saliva mixed in with the milk at sucking time, as well as having an effect on some of the digestive enzymes involved. However it appears that the functioning of the oesophageal groove is not affected by the method of feeding and few clear advantages have been demonstrated for teat feeding. The choice therefore boils down to managerial convenience.

The process of calf rearing has been mechanised so that warm milk replacer is available to the calf through a teat at ad libitum or restricted levels. The machines can reduce labour but experience shows that they require skilful supervision; rigorous attention to size of group and the unthrifty calf are essential.

Calves should be given access to clean water, small quantities of fresh calf starter and to good-quality hay, from an early age. Because calves cannot consume significant quantities of any of these three until they are at least two weeks of age it may be advisable to wait until then before providing the calf starter. This will not only cut down on wasteful spoiling of expensive material but may provide a stimulus to the start of consumption at a stage when the calf is physiologically ready to commence eating solid feed. This is because the calf's curiosity is aroused by new objects in its pen. This assumes that the calves are penned individually and do have access to some bedding material in the first fortnight of life.

Where bucket feeding is practised, individual penning is an advantage in that calves are prevented from developing vices such as navel and ear sucking and their progress is more easily monitored. The usual individual pens still allow the calf to communicate with its neighbours and performance is improved under these conditions.

Weaning

It is theoretically possible to wean most calves from about four weeks of age onwards but the best compromise between calf performance and cost of rearing has been shown to be around five to six weeks of age except for specialised production systems like veal production.

The criteria for weaning stage are as follows:

1. Age.
2. Body weight.
3. Intake of solid feed.

The optimum level of these three criteria may vary with breed and with the production system. In theory, where calves are individually penned, the level of dry concentrate consumption is the best criterion for weaning and some calf rearers set targets like 0.5 kg per day for weaning. In practice it is easier to manage a large calf rearing system on a 'weaning according to age' basis, provided flexibility is exercised so that the weaning of backward calves can be delayed according to need. Liveweight is not a good criterion since calves vary substantially in birthweight so that some calves are ready for weaning at a weight when another calf is born.

Calf rearers who buy in calves from unknown backgrounds are in a particularly difficult position since they often do not know the date of birth of the calves and their previous feeding history. In such cases all bought-in calves should be put on good-quality milk replacer for at least three weeks unless their concentrate consumption is monitored and clearly indicates that they are ready for weaning.

Weaning of calves may sometimes be profitably delayed if cheap sources of skim milk or butter milk are available. These can be introduced at three weeks of age and form the main liquid feed after a transition period of ten to fourteen days.

Weaning may be carried out as a gradual process over four to seven days or it may be carried out abruptly. Both systems have their advocates but it seems that, provided the right stage for weaning is clearly judged, the speed with which it is achieved matters less.

After weaning calves should be kept in their rearing pens and on the same calf starter as they are used to. After a 2–3 week period they are then usually moved in groups to their growing accommodation and the diet gradually changed to that appropriate for the young growing calf.

References

PRESTON, T. R. and WILLIS, M. B. (1975), *Intensive Beef Production,* Pergamon Press.

REID, T. J. et al. (1963), 'Effects of plane of nutrition during rearing on lifetime performance of dairy cattle', *Proc. 1963,* Cornell Nutrition Conference.

ROY, J. H. B. (1970), *The Calf*, Vol. 2. 'Nutrition and health', Iliffe Books Ltd.

ROY, J. H. B. (undated), 'The composition of milk substitute diets and the nutrient requirements of the pre-ruminant calf', Roche Information Service.

SHUJAA, T. A. (1989), *Studies of the effects of probiotics on the health and performance of ruminants*, PhD thesis, University of Wales, Bangor, pp. 196.

Chapter 7

FEEDING PRACTICES—INDOORS

Feeding theory must be applied in the practice of feeding systems for cattle. The management system and the available feed dictate the details of the system and how the principles are put together to good effect.

IN THE OPENING chapter it was shown that all cattle feeding systems involve some substantial extra feeding, other than grazing, during part of the year. Although this is sometimes achieved by supplementation of pasture in the most favourable grass growing areas, in most others it involves a period when the cattle are fed away from pasture. In drier areas the cattle can be kept outside in feedlots or yards; in others where the weather is too wet, too cold or too hot, some form of shelter or housing is employed.

FORAGES

The basis of cattle feeding is the forage. Circumstances may however dictate either a minimum of forage, as in some semi-arid areas, or a maximum, where forage cost is low, as in some of the moist temperate areas. Whatever the circumstance, forage quality and forage utilisation have to be closely watched.

There are several types of forage in use, as follows.

Fresh-cut forages. These can include grasses, cereals, legumes (particularly lucerne or alfalfa), brassicae (kale and cabbage) and many others. Some of these species are suitable for repeated cutting over a long growing season either with or without irrigation. To ease the problems of harvesting and transport, the material is usually cut at a more mature stage than when it is grazed directly by the animal. Also in order to avoid wastage, allowances are reduced, thus encouraging the animal to consume a higher proportion of the stemmy, less nutritious material than when allowed full rein to its selectivity as under grazing conditions.

One feature of fresh-cut forage as compared with ensiled forage is that, when comparing similar original material, the fresh material will not suffer some of the negative intake characteristics associated with silage. There is also evidence that the protein value is reduced in ensiled material resulting in a significant reduction in the UDP fraction.

Dried forages. The most common traditional way of conserving forages is by drying naturally in the field. The resulting material—hay—is widely used for cattle feeding. Artificial drying of forage is less common because of the costs of drying but artificially dried, high-quality grass or lucerne/alfalfa, may be incorporated into cattle diets as a minor component because of its rich supply of carotene, minerals (notably calcium) and its high-quality digestible fibre.

Hay is a variable material because of the variation in stage of cutting in relation to plant maturity and because of the varying losses in the field curing process. The optimum stage to cut forage for hay is a compromise between early maturity and therefore higher nutritive value, and the yield obtained per cut and per hectare, coupled with the ease of drying. In wetter areas hay cut at an early stage of growth is highly vulnerable to weather damage because of the long drying period required. Furthermore even in dry areas young material is prone to loss through overdrying of the leafy parts and their shattering and loss.

Many studies have shown that, provided the same original material is compared, well-made hay has some advantages over well-made silage. In particular there can be differences in voluntary dry matter intake as shown in Chapter 3.

Some authorities (ARC 1980) however claim that these differing intake characteristics can be minimised by good silage-making practice.

Silage. Many forages can be preserved by the process of ensilage. In this process of controlled fermentation, carbohydrates in the material are fermented into a predominantly lactic acid end-product under the action of lactobacilli. The material, under the appropriate conditions, reaches a high enough level of acidity (pH 3.8–4.4) quickly to prevent the action of other micro-organisms which would cause putrefaction. Factors conducive to good fermentation include cutting material when it is dry, when it contains a high level of soluble carbohydrates and to subject it to a quick process of heavy consolidation to exclude air.

The advantages of silage as a way of preservation is that it eliminates the need for the field curing process required for hay and cuts down the vulnerability of the process to wet weather. It also increases the flexibility of stage of cutting since ensilage is technically possible from a very young stage of growth to a mature stage. This flexibility is enhanced when silage additives are used in the conservation process. The most successful additives so far in use are those where acid is added to the material during the harvesting process to encourage the correct fermentation.

The feeding of silage has been enormously eased by the development of handling machinery and, for the smaller family farm, the use of self-feeding from silage clamps marked one of the major agricultural advances of the present century.

As a feed material the possible negative intake characteristics and its low UDP value have already been noted.

Forage Nutritive Value

Although it is difficult to generalise about a class of feedingstuff as variable as forage there are several aspects of its value that are important in cattle feeding.

A major decision the cattle feeder often has to make is what forage nutritive value to aim at. By choice of cutting date the farmer has the possibility to produce material from the highest to the lowest quality. Fig. 7.1 shows the effect of date of cutting on the quality of grass silage under good land quality conditions in the U.K. Similar figures can be produced

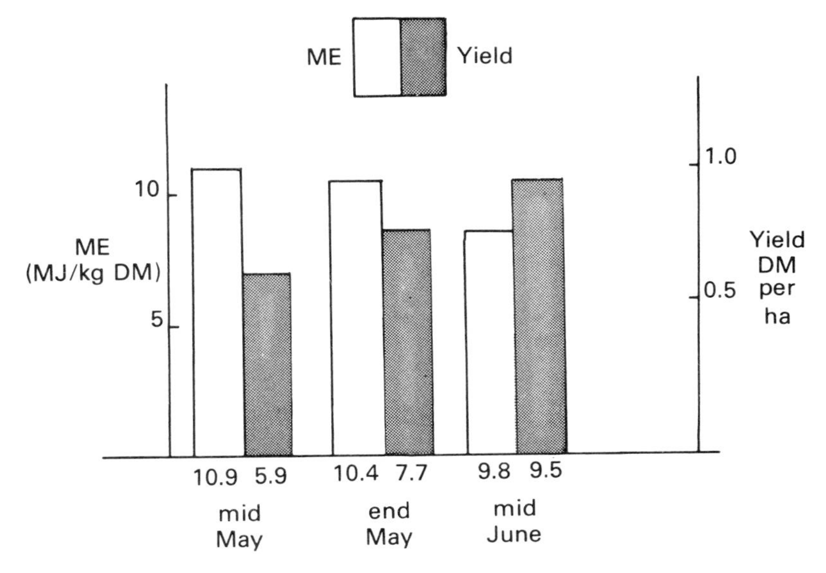

Figure 7.1. Cutting date in relation to first cut silage quality in the U.K. (mainly ryegrass sward)

for varying types of material in any part of the world.

As stage of maturity advances several consequences follow:

(a) Forage nutritive quality decreases
- Digestibility i.e. ME value falls
- Total crude protein falls
- UDP falls
- Crude fibre increases

(b) Forage intake characteristics decrease
(c) DM content and yield per hectare increases
(d) Cost of harvesting and material handling per tonne falls.

The optimum stage at which to cut the forage is therefore a complex one depending on:

- type of enterprise e.g. dairy or beef
- cost of concentrates relative to forage costs
- the availability of land.

An example of the type of calculation that can be done is shown in Table 7.1 where the consequences of two stages of cutting for silage are examined in relation to a beef-finishing enterprise. This type of calculation usually shows that silage should be cut neither very young nor very mature but at an intermediate stage which gives the best compromise between bulk and quality. In the example in Table 7.1 the slightly greater area required with medium-quality silage is offset by extra grazing and lower cost.

Table 7.1. Silage, cereal and land requirement for feeding 100 beef cattle of 400 kg liveweight during a 150-day winter feeding period to achieve the same daily gain on a high- (10.9 ME) and a medium- (10.2 ME) quality silage

	Medium-quality silage (10.2 MJ/kg DM)	*High-quality silage (10.9 MJ/kg DM)*
Number of cuts	2	4
Last cut	end July	mid August
Yield of silage DM (t/ha)	9.0	8.5
Average silage DM consumed per beast/day	6.5	8.5
Average barley DM consumed per beast/day	2.0	—
Total fresh silage required per winter @ 25% DM (t)	390	510
Total barley required per winter @ 85% DM (t)	35	—
Area of land required for silage (ha)	10.8	15.0
Area of land required for barley @ 5t/ha (ha)	7.0	—

Because so many factors are involved it has been difficult in the past go give a precise estimate of the optimum forage quality applicable to any individual farm. However the recent development of computer programmes gives promise that such procedures will soon become routine farm advisory/extension tools.

For many situations, including milk production and beef rearing and finishing systems, most producers making grass silage will not be far from the optimum if they aim for material that is well made and with an energy ME value in the range 10.0–10.5 MJ/kg DM.

Forage Rootcrops

Although not easily classed as forages, since in many ways they are really concentrated feeds, rootcrops have often some of the same problems as those involved in forage feeding, due to their bulky nature.

Rootcrops include a wide variety of crops that have been traditionally an important element in the diet of cattle in the wetter, moister regions of Northern Europe. They include swedes, mangolds and fodder beet but many vegetables used primarily for direct human consumption such as carrots and potatoes are often used in time of glut. Sometimes also the by-products of cash crops such as sugarbeet tops can be included in this category.

The rootcrops are characterised by a low dry matter content in the range 80–250 g/kg DM, but, unlike grass, they are usually stored without processing either in the ground as they have grown or in covered clamps.

For cattle feeding most of these materials have to be processed so that they are cut into smaller pieces; this is essential for the harder, high dry-matter roots such as fodder beet and for potatoes if the danger of choking is to be avoided.

The nutritive value of the roots, based on their fresh weight, depends mainly on their dry-matter content; variation in the value of the dry matter itself is less, e.g. ME is usually in the range 11–13 MJ/kg DM and UDP 15–25 g/kg DM. They are all materials that would be classed as concentrates in terms of their DM content, with crude fibre ranging from about 40 g/kg DM (potatoes) to about 110 (turnips). The main value of roots is that they are capable of producing amongst the highest yields per hectare of high value, easily storable winter feed energy. Their main drawback is the labour associated with their growing, harvesting and feeding.

Forage Feeding Methods

These methods include the whole range from the traditional methods associated with the cow stall system to the most sophisticated mechanised methods available to the present-day farmer.

In the traditional cow stall, feeding was and often still is associated with hay, roots and a small quantity of grain concentrates. Hay barns were usually constructed abutting on to the cow accommodation so that it could easily be carried through to the cattle either in wads of cut loose hay or nowadays more commonly in bales. The system is also adapted to carrying in cut roots by the barrow load from an adjoining root preparation room. Such a system is reasonably easily adapted to the carrying of silage, cut in wads from a clamp and wheeled in like the roots. Roots, concentrates and silage are usually fed in shallow troughs or on the concrete floor. Hay may be fed in the same way but it is also often fed in racks from which the cattle pull the hay wisp by wisp.

A major development in cattle housing that influenced forage feeding methods was the development in arable areas of large open cattle yards or courts in which the cattle were not tied but wandered freely on a liberally strawed area. This enabled more flexibility in the placement of hay racks and even enabled movable racks and troughs to be introduced which could be moved as the forage was consumed.

Self-feeding of Silage

When silage-making became widespread, particularly after the 1940s, the system of placing silage clamps next to cattle yards and allowing self-feeding became common and it is still a common method of feeding silage in the United Kingdom at present (Plate 9). In this system cattle are allowed as near 24-hour access as possible to the silage face and they self-feed ad lib. Constant access means that the length of feeding face per beast may be reduced compared with that necessary on a restricted system where all the cattle have to be able to eat at the same time. Concentrates are then fed in the parlour or in separate troughs.

Mechanised Silage Feeding

In spite of its simplicity and cheapness, self-feeding of silage can set problems for the larger unit where groups of cattle have to be fed separately. Mechanised methods have therefore been developed to move the silage from the clamp face to the area where the cattle are to eat it (Plate 10). Also in

Plate 9. (*above*) Cows on self-fed silage (*Farmers Weekly*)

Plate 10. (*below*) Mechanised indoor feeding of cattle (*Farmers Weekly*)

Plate 11. (*above*) Mixer wagon delivering feed through the side delivery mechanism

(*left*) A view of the paddle mechanism, which provides simple and effective mixing of ingredients

many areas making and storing silage in towers is popular. Under both circumstances the silage may be moved from the storage to the feeding area using a forage waggon which is usually equipped with a self-unloading device so that the forage is delivered neatly at a feeding barrier. More expensive and sophisticated machinery may incorporate automatic weighing devices, usually based on load cells, and mixing augers or paddles so that a variety of forage ingredients may be mixed together and with concentrates as shown in Plate 11. This is a flexible system which allows the feeding of large numbers of cattle with relatively little capital outlay. The system is most fully developed in the large feedlots such as those of Arizona in the United States, where many thousands of cattle can easily be fed with a small labour force.

Another less flexible system of mechanised feeding is that more commonly associated with the tower storage of silage. Here silage is automatically unloaded from the top of the tower and delivered to a conveyor system which automatically conveys the forage along the troughs for the cattle. Sometimes these systems allow for mixing so that at least loosely mixed material is conveyed along the system.

Forage Processing

Mention has already been made of the treatment applied to forage roots to make them amenable for feeding to cattle. Grass and legume based forage does not need to be treated to be eaten by cattle and most hay, for instance, is consumed by cattle in the long form.

Chopping and grinding. There is no nutritional advantage in chopping the forage material into smaller particle size but it is sometimes convenient or essential for some forms of feeding. There are some advantages in chopping material for silage making because it is more easily consolidated and can often lead to better-made silage.

It may also be more available to the cattle on a self-feeding system, particularly benefiting young and timid cattle whose access to the silage face may be limited by the larger, more aggressive cattle.

Chopping is of greatest relevance where mixed diets are made. Chopping material into short 10 mm even length is invaluable in the aim of achieving a uniform blend which is not amenable to selective eating by the cattle. It is now well known that the over-processing of forage can have a deleterious effect on its nutritive value to cattle. Material that is finely ground passes through the rumen more quickly and can therefore often be consumed in greater quantities but its digestibility is usually less (Table 2.3). Overall the feeding value of low-grade forage may be improved by grinding but not that of better quality forage. Where forage is valued for its 'roughage' value, particularly in dairy cow diets where forage is at a premium, then grinding would be avoided, since it can only exacerbate a situation where the cow is already marginal in its roughage supply and lactation efficiency, and butterfat content is probably already being affected.

Whilst low grade precision chopping (down to particle size about 1 cm) is widely adopted in mechanised feeding for the reasons stated, the use of fine grinding of low-grade material for use in beef cattle finishing is not widely practised because of the energy required to grind the material and the relatively small gains in performance since the advantage of higher intake is diminished by the depression of digestibility.

Chemical treatment of forage. An alternative to fine grinding is the use of chemical agents, notably alkalis, in the treatment of forage. Again, as with grinding, low-grade material like cereal straws is more affected by the treatment than better-quality material. As shown in Table 3.5, chemical treatment with an alkali like sodium hydroxide acts mainly by increasing the digestibility of the material with associated effect on DM intake. Because both effects are favourable and additive, significant and important improvements in feeding quality can be achieved by alkali treatment and it is technically feasible to upgrade poor straws and hays to medium-quality forage. Whether it is always profitable to do so is not so easy to answer. Since the biggest improvements in digestibility are obtained with the lowest-quality forage to start with, it is usually only economic to apply the treatment to material like straw. These materials, even when upgraded, are only moderate-quality forage and it may only be possible to incorporate them to a limited extent in the feeding of high-producing stock. There is also some element of danger in handling alkalis, which can only be overcome by fairly expensive capital outlay. Another unfortunate feature, from the recent work done on this problem, is the rather disappointing variability in the results, which are not, so far, fully explainable. It is possible that many of the snags can be ironed out with more research and that a real advance in animal production may be achieved. At present it is premature to hope that all the tonnes of cereal straw burnt at harvest time will provide useful and profitable cattle food.

CONCENTRATES

The concentrate fraction of cattle diets is so described because it is a relatively concentrated energy source but in reality there is a very wide variation in its ability to supply nutrients. Some of the concentrates, although still a concentrated energy source, are primarily sought for their protein value, particularly UDP that may be deficient in the forage source. As shown in Table 4.2 concentrates include cereal grains, animal by-products and oil seed residues amongst other materials. They can also include high-quality artificially dried grass which could be considered to be at the margin when feed is categorised into either forage or concentrates. Many concentrates, particularly the plant by-products, are of relatively high crude fibre content, e.g. sugar beet pulp, undecorticated cotton seed cake, brewers' and distillers' grains. Although these materials may be at the lower end of the energy value scale, they provide an important source of variety, particularly in formulating dairy cow diets, where roughage can be a crucial element.

Recently there has been an increasing use of various high-fat materials to enrich the lipid content of the diet and to provide an easy way of manipulating energy content without making other drastic changes.

Concentrates as a whole are good sources of phosphorus and complement forages which are normally rich sources of calcium. Some of the animal by-products such as fishmeal are exceptions in being a rich source of many important minerals, especially calcium and phosphorus.

Concentrate Processing

Many of the concentrates, for example, cereal grains, need some processing before they can be fully digested by cattle. This is because of the kernel coat or husk—the cuticle—which prevents full access by digestive juices to the kernel itself. Many experiments have shown that rolled or bruised cereal grains, where the grain has been squeezed between two rollers to rupture the husk, when fed to cattle shows something like 10 per cent higher digestibility than whole grain. An example of an experiment carried out in Aberdeen in the 1960s is shown in Table 3.6.

On the other hand the point made about fine grinding in relation to forage is equally valid for grains in that overprocessing reduces the already low roughage value and can lead to digestive problems in cattle given high-grain diets.

Alkali processing of concentrates is not valuable since high-quality material is hardly improved by the treatment (cf. low-grade forage). However some interest has recently been shown in the possibility of treating grains in order to protect the protein so that the UDP value is improved. The use of formaldehyde and certain tanning agents can achieve a protective effect by chemical reaction with the surface of the concentrate. However it has been shown that the treatment may prove ineffective due to overprotection leading to a reduction in the overall digestibility. In spite of some promising results with the 'protection' of materials like soya bean, there is as yet no convincing evidence that reliable profitable improvements are invariably gained in this way.

ALLOCATION OF CONCENTRATES AND FORAGE FOR DAIRY COWS

One matter which was not discussed in detail in the chapter on diet formulation (Chapter 5) was that of the allocation of concentrates and forages within a global total for a group of cattle for a period of time such as the winter feeding period. Modern feed rationing systems grew up with the assumption that the main object of the exercise was to meet needs or requirements of animals. This assumption has been rejected in Chapter 5 in favour of a more direct objective, that of formulating diets that maximise the farmer's profit. This objective can be tackled in two stages:

1. Assessment of the optimum forage and concentrate allowance over a clearly defined period of time (as in Chapter 5).
2. Determination of the optimum method of allocating these total amounts over the period in the most effective way.

The first stage has already been discussed in some detail in Chapter 5. It is therefore necessary to look closer at Stage 2. Because it is technically feasible, under conventional feeding methods, to allocate concentrates to dairy cows individually, it is possible to allocate concentrates to cows over the full range encompassed by feeding individually according to yield at the one extreme or at the other by feeding all cows within the group a fixed daily allowance for the whole period.

Arising out of the traditional emphasis on feeding to meet requirements, it has seemed to be a natural and unchallengeable consequence that feeding according to yield is a good thing, i.e. the most efficient way of feeding cows. Little attention seems to have been paid to the apparent inconsistency of this approach with that adopted for most other stock where individual feeding is less easily practised. For these stock, in spite of known substantial changes in body composition, for example, it has not been considered important even to adjust diets frequently let alone try to cater for differences between individuals.

In recent years several experiments have been carried out to examine the optimum way of allocating concentrates to dairy cows. These experiments have included those where cows have been given complete diets matched to stage of lactation as compared to those on a single standard formulation. Others have made a direct comparison of possible allocation methods. These have included the classic experiment of Østergaard in Denmark and those of Gordon in Northern Ireland. Taken overall, the results so far show that different methods of allocation have not shown a major influence on the efficiency of feed use even when cows of different yield potential and different stages of lactation have been examined.

Some examples of the results of the major experiments are shown in Table 7.2.

Table 7.2. The effects of different methods of allocation of concentrates, in conjunction with ad libitum silage, to dairy cows over the lactation. Results of some Northern European studies

Reference	*Period of lactation when the concentrates were fed (post calving)*	*Types of distribution system compared*	*Results*
Østergaard (1979) (Denmark)	36 weeks	Various, including flat rate	No important effect discerned on lactation milk yield, FCM yield, milk quality
Gordon *et al.* (1981) (Northern Ireland)			
spring calving	2–3 months	Step feeding and flat rate	
autumn calving	5 months	Step feeding and flat rate	
Johnson (1983) (England)	20 weeks	Step feeding and flat rate	
Taylor & Leaver (1982) (Scotland)	20 weeks	Step feeding, flat rate and flat rate based on initial yield	
Kroll, Owen and Whitaker (1986) (Wales)	6 months	Complete diet uniform and stepped composition	

Their findings can be summarised as follows:

For dairy cows given a basal diet of ad lib medium- to good-quality silage in conjunction with concentrates or given ad lib access to a complete diet containing such silage, there is no evidence so far of any major effect of method of concentrate allocation. Thus systems where one complete diet or one flat rate of concentrates is given for the major part of the lactation seem to be just as efficient as other more sophisticated systems like feeding according to yield.

This is a major watershed in dairy cattle feeding systems

Easy-feed forage is still the mainstay of European cattle feeding in winter *(Farmers Weekly)*

since it not only has immediate practical repercussions in terms of the flexibility of feed allocation systems but, like all revelations of the working of nature, it has profound repercussions on thinking and the course of future development in this field.

In terms of thinking it emphasises very clearly that cows are capable of responding to better feeding at all stages of lactation, not just in the early part. It is also apparent that cows at all yield levels are capable of responding to better feeding. A classic but rather neglected experiment which emphasises some of these points is that carried out in Denmark in the 1950s where a group of ordinary Red Danish cows were subjected to high feeding and management after a fairly ordinary previous lactation on commercial farms. The results are summarised in Table 7.3. They show that all cows were able to respond to better feeding and management, both in milk yield and in milk quality.

Table 7.3. The response of commercial Red Danish dairy cows to high management (including 4 × per day milking) and to careful heavy feeding irrespective of stage of lactation or yield

	Pre-experimental years		*Experimental year*
	1945–6	*1946–7*	*1947–8*
Days in milk	328	287	365
Milk production (kg)	4,372	4,567	10,526
Fat (%)	3.73	3.69	3.93
Butterfat yield (kg)	163	169	414

Source: Larsen and Eskedal (1952), '260 beretning fra forsøgs laboratoriet Copenhagen'.

A finding of this kind, which seems contrary to much of the basis of cow feeding in the last half-century, is only slowly accepted and is likely to be subject to close testing in future. The fact that method of concentrate allocation has so far been shown to be relatively unimportant only on systems based on medium- to good-quality silage and where either the forage or the whole diet is ad lib, is a minor qualification since systems of cattle feeding in the foreseeable future are likely to encompass both. So far there is little evidence one way or the other as to what would happen when silage quality is lower or it is given on a restricted basis since much of the cow experimentation in the past has been on a short-term, part-lactation basis, unsuitable for testing whole-lactation effects.

Specific Disadvantages of Heavy Concentrate Feeding in Early Lactation

If the evidence on the effects of systems of concentrate allocation is accepted, the notion that cows should be fed heavily with concentrates in the first few weeks of lactation, to encourage a high peak yield, is not tenable. In any case it was always a fallacy to step from the observation that variations in lactation yield are closely associated with the level of peak yield, to the generalisation that cows should therefore be heavily fed to achieve the maximum peak yield. Just because ability to run a marathon is associated (negatively) with body weight it does not follow that if you fast for a month before an event you will necessarily do much better.

There are some specific problems, some of which are dealt with more fully later, associated with heavy concentrate feeding in early lactation. These include ailments such as ketosis, abomasal displacement, laminitis and the mal-partition syndrome involving low fat milk and reduced lactation efficiency. Cows of relatively limited intake capacity will eat concentrates at the expense of their necessity for roughage (see Fig. 7.2).

Taking specific dairy cow systems into consideration the evidence so far available on these is in line with the general statement given above.

Spring-calving herds. These herds calve normally one to three months before the cattle are turned out to grass in the spring. It is apparent that in these herds not only is method of concentrate allocation unimportant in deciding lactation performance but the level of concentrate supplementation

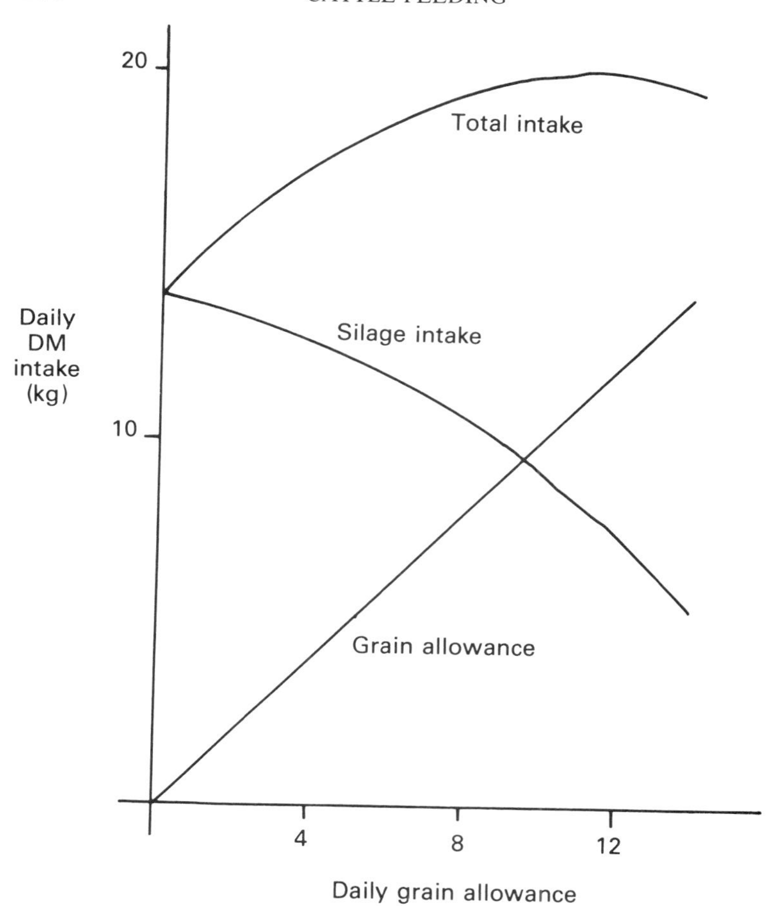

Figure 7.2. Daily intake of a cow (550 kg) given silage ad libitum and various level of grain

Source: Østergaard, V. (1979), 482. Beretning fra Statens Husdybrugs Forsøg, Copenhagen

before turn out also has a lesser effect on performance within limits (Fig. 7.3).

At the other end of the lactation on the other hand which, for the spring calving herd, coincides with the autumn period and the transition to housing in early winter, economic

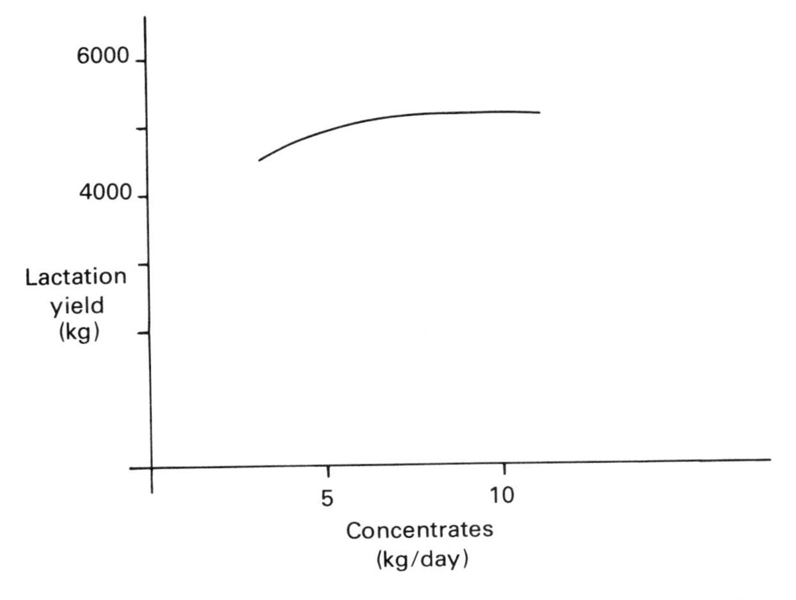

Figure 7.3. Response of spring calving cows to daily concentrate intake before turnout
Source: Gordon *et al.,* 1981

benefits are apparent for concentrate supplementation as shown in some recent Trawsgoed EHF (Wales) results.

Autumn-calving herds. These herds calve at the end of the grazing season—Sept./Oct./Nov. in the U.K.—and all available pertinent evidence does not show a significant response to differing methods of concentrate allocation as far as total lactation performance goes.

Non-grazing herds. For cows that are fed the year round on cut forage and concentrates the evidence indicates that at least up to 250 days of lactation method of concentrate allocation does not have an important effect on performance. Østergaard's Danish experiments looked at the first thirty-six weeks (252 days) of lactation in the results discussed in Table 7.2.

Concentrate Allocation Systems in Dairy Herds

Having established that cow performance can be maintained on a variety of concentrate allocation systems the systems in use and their pros and cons can be considered.

Feeding According to Yield

With the advent of the self-feeding of silage a relatively simple system of feeding dairy cows became widely practised. This involved allowing the cows ad lib access to their forage and feeding concentrates at milking time in the parlour, usually twice a day. The system involved making an assessment of the milk that a cow was likely to produce on the silage alone, e.g. 10–15 kg per day, and then feeding according to the yield given in excess of this, usually at a rate averaging in round terms about 0.4 kg concentrates per kg of milk. Thus a cow giving 30 kg per day, in a herd where silage was assessed at a 15 kg milk production basal level would be given 6 kg of concentrates per day in two feeds.

It was then thought that feeding in early lactation was more important than at other times and the notion of 'lead' or 'challenge' feeding arose where cows were given more than the average (say 0.4 kg/litre) in early lactation to encourage high peak yield. In view of the results shown earlier in this chapter such variations on the simple theme of feeding according to yield do not appear well based and can exacerbate the ill-effects of overloading the cow with concentrates in early lactation. These variations will not therefore be discussed further.

As yields have increased, the system of feeding according to yield has had to be modified to try to reduce the large quantity of concentrates given twice a day at milking time. Cows find it difficult to consume the concentrates fast enough in the time in which they are in the parlour for milking and it is known that large infrequent feeds can lead to digestive upsets as well as incurring the penalties in terms of conversion efficiency dealt with earlier.

Feeding according to yield has some advantages, the primary one being that there is no need on this system, in its simplest form, to provide any trough space, other than that

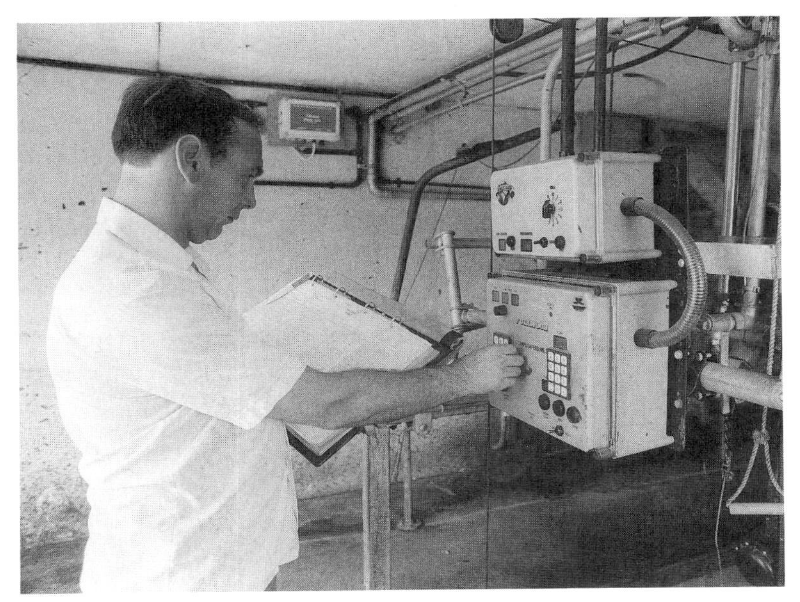

Plate 12. Computer feeding of concentrates in the milking parlour

associated with the milking stall (Plate 12). Another advantage, particularly to the inexperienced feeder starting in a new situation, is that it rigidly limits the amount of concentrates given. As the operator gets to understand better the potential of the cows in his system this advantage largely disappears.

The main disadvantage of the system is that it involves regular recording of the milk output of individual cows, which might not be justifiable for any other purpose. It also requires some mental concentration to identify cows, check up on the allowance required and dispense this. To overcome some of this disadvantage sophisticated parlour equipment is available as shown in Plate 12. Such equipment is expensive and at present not fully developed in terms of reliability.

Another disadvantage already referred to is that for twice-daily milking most farmers find that the two parlour feeds need to be supplemented by some feeding outside the parlour, thus adopting a different mixed system dealt with later.

Frequent Feeding of Concentrates

Some of the ill effects of heavy concentrate feeding on low forage:concentrate ratios can be mitigated by splitting up the concentrate allowance into several smaller meals spread more evenly over the twenty-four hours. By this means digestive upsets are avoided, protein is more efficiently utilised and lactation partition is more normal, e.g. milk fat content is improved. Table 7.4 illustrates some of the work done on this point.

Table 7.4. Effect of concentrate feeding frequency on the lactation performance of dairy cows

	Level of protein in concentrate					
	Low (128 g/kg DM)			*High (202 g/kg DM)*		
No. of concentrate feeds per 24 hrs	2	4	22	2	4	22
Daily concentrate DM intake/cow (kg)	6.0	5.9	6.1	6.1	6.1	6.1
Daily DM intake (kg)	13.9	14.1	14.0	14.4	14.6	14.6
Milk yield (kg/day)	15.3	15.2	15.9	17.3	17.7	17.2
FCM (4%) yield (kg/day)	17.2	17.6	19.1	18.8	19.4	19.4
Milk composition (g/kg milk):						
Fat	45.3	45.9	48.3	43.1	43.6	45.8
S.N.F.	87.7	88.1	88.5	88.0	88.4	89.1

Source: Gill and Castle (1983), *Animal Production 36,* 79–85.

Frequent feeding of concentrates can, in practice, be achieved by several of the concentrate allocation methods that are discussed in the following section including a combination of one of them with parlour feeding as mentioned above.

Out-of-parlour Concentrate Dispensers

A fairly recent development is that of allowing cows access to concentrate feeders away from the milking parlour (Plate 13). The history of this development goes back some time. For instance in the 1950s one ingenious American worker constructed a feeder which was coupled to a water bowl and a cow was dispensed concentrates according to the amount of

Plate 13. Out of parlour concentrate feeder (*Farmers Weekly*)

water consumed. This was a plausible idea since it is well known that water intake is associated with milk yield and feed intake. However the idea does not seem to have caught on, probably because cows may well be able to distort their drinking habits to defeat the plans of mice and men.

The first more recent development of the out-of-parlour concentrate dispenser arose from the possibility of magnetically or electronically controlled feeder access. Broadbent at Aberdeen, for example, developed the selective access feeders or doors where certain cows could be given a chain suspending a transponder which opened the doors. The idea behind this and the magnetic type was to allow high-yielding cows access to an ad lib supply of concentrates. This system can only succeed however if the feed to which selective access is given is one which can be consumed in large quantities efficiently. Concentrates on their own are not in this category so some kind of complete diet incorporating chopped hay or straw would be more suitable.

More sophisticated 'programmed' dispensers are now available which can limit the amount available per feed and the interval between feeds for any individual cow. The more expensive models are capable of varying the amounts and intervals according to the yield group of the cows.

These dispensers allow the extension of the older principle of individual feeding according to yield to modern high-yielding conditions. They retain the advantage of keeping feeding space down but at a considerable increase in cost.

Flat-rate Feeding

Arising out of the newer findings about cattle feeding, described earlier in the chapter, a system of feeding called flat rate or level feeding has developed. Some have tended to associate flat rate with a specific rather low level of feeding concentrates but there is no reason why the system should not be applied to any overall level of concentrate feeding.

The essential of the system involves a decision on the appropriate concentrate allowance per cow as in Table 5.7. This is then applied to the herd as a whole for the period in question, depending on whether it is a spring-calving, autumn-calving or other kind of herd.

Those who do not fully accept that concentrate allocation is not important may wish to modify the system by splitting their herd into two or more groups or by adjusting concentrate allowance to a group as lactation progresses. This may often be possible without any extra complication of the system.

Flat-rate feeding can be achieved in practice either by parlour feeding or by feeding in troughs outside the parlour or by a combination of both. This system is simple and effective and its only apparent disadvantage is that trough space has to be provided for all cows in the herd to feed at the same time as its group mates. The term 'flat rate' as applied to the group may not literally mean the same consumption per cow since cows vary in their rate of eating.

The Complete-diet System

In this system all the ingredients of the diet, both forage and concentrates, are thoroughly mixed to provide a uniform blend from which cows cannot significantly exercise differential selection. Ingredients of the diet have to be processed so that particle size is small enough to achieve the above conditions and the diet is made available continuously for most of the 24 hour period.

The system can be operated without any other feeding of concentrates, for instance in the milking parlour. The diet can be formulated to cater for the needs of all lactating cattle for the major part of the lactation. It is convenient and possibly cost effective to provide dry cows and cows in the last few weeks of lactation with a different diet, such as one consisting solely of medium-quality silage.

Some practitioners of the system do feed a small concentrate allowance in the milking parlour and many operate a system with two or more 'lactation' diets. So far there is no experimental evidence to support the value of such practices but if such modifications can be achieved without losing the basic simplicity of the system then there need be no significant disadvantage.

Other dairy farmers practise a system which is a combination of the complete-diet system and feeding according to yield. This entails the cows having access to a basic, relatively low-concentrate mix and then being given concentrates according to yield in the milking parlour. The advantage of this variation over the traditional system of feeding according to yield is that the inclusion of a basal level of concentrates with the forage cuts down the allowance needed in the parlour and makes it more feasible for high-yielding cows to consume their allowance at milking time.

The complete-diet system is a cheap and effective system of exercising tight control in the feeding of a large dairy herd. Since cows on this system are forced to eat a fixed quantity of forage for each kilogram of concentrates consumed the system is a safe one and avoids the large variation in forage consumption that occurs with separate feeding of concentrates. This safety factor, allied with the more frequent consumption of smaller quantities of concentrates, explain the generally improved milk quality experienced by those on this system. Improvements of 0.15 percentage units of butterfat and 0.05 of solids not fat have commonly been observed in comparisons of complete diet with separate feeding of forage and concentrates.

Table 7.5. Comparison of a complete diet* given *ad libitum* with the same ingredients given separately (with forage *ad libitum*)

	Complete diet ad libitum	*Ingredients given separately*
Mean daily DM intake (kg)	16.5	14.3
Mean daily milk yield (kg)	23.0	21.6
Milk fat content (%)	0.5 units greater on complete diet	
Rumen fermentation (Ratio of lipogenic to non-lipogenic end products)	improved on complete diet	

* Diet consisted of 60% concentrates and 40% forage and contained 20%, 20%, 10% and 50% on a DM basis of maize silage, lucerne silage, sugar beet pulp and dairy concentrates respectively.

Source: Phipps, R. H. Bines, J. A. and Weller, R. F. (1981), *Annual Report*, National Institute for Research and Dairying, 1981.

Some large-scale trials at Reading have yielded some interesting results on the comparison of complete diets and separate feeding, as shown in Table 7.5. The indications are that whilst at medium values within the range of forage:concentrate ratio, the complete-diet system is little superior to separate feeding, there appear to be more substantial benefits from the complete-diet system at very high or very low concentrate levels.

Beef Cows

Feeding Beef Cattle

Many of the general points made about dairy cows are applicable to the feeding of beef cows and also of growing finishing beef animals. Generally speaking beef cows can be fed on all the systems outlined except of course that parlour feeding is not relevant. Beef cows are usually expected to make full use of summer grazing and it is more common than with dairy cows to exploit the compensatory mechanism to the full, whereby cows are fed below optimum under winter housing conditions and use up their body reserves to be replenished during the grazing period.

This is reflected in the higher optimum forage:concentrate ratio for beef cows (Table 5.8).

Growing/finishing cattle. Some experimental work is available on the comparison of complete mixed diets with separate feeding. Work at the School of Agriculture, Aberdeen is summarised in Table 7.6.

Again as with dairy cows mixed diets seem to be of the greatest value when very high or very low levels of concentrates are being fed.

Housing and Machinery Aspects

Having covered the systems used for the feeding of cattle it is pertinent to describe some of the physical features that ensure efficient feeding. There are several principles involved:

Table 7.6. A comparison of silage and barley fed separately or as a mixed complete diet for beef cattle

	Proportion of barley DM in the diet			
	0.2	*0.4*	*0.6*	*0.8*
Daily DM intake (kg):				
separate	7.3	8.1	7.5	7.3
complete mix	7.9	8.6	8.6	7.9
Daily gain (kg):				
separate	0.68	0.90	0.81	0.83
complete mix	0.80	0.85	0.83	0.88

Source: Petchey, A. M. & Broadbent, P. J. (1980), *Animal Production 31,* 251–7.

1. Group size—the smaller the better for the animal;
2. Trough space—adequacy will prevent the ill-effects of pushing and bullying;
3. Prevention of feed waste.

Group size. Wherever possible, group size should be minimised. Optimum group size is a function of the space allowance. Larger groups can be tolerated where space allowance is liberal since this can allow the formation of subgroups. Where space allowance is restricted as for instance under some of the slatted floor systems, group size should be kept small.

Although it is difficult to generalise, several investigations have noted that under commonly found housing conditions animals tend to suffer if kept in groups larger than about a hundred. This may be because under normal space allowance it is difficult for animals in groups above this size to get to know each other and establish their position in the hierarchy. Often it is convenient to have smaller groups and this is to be encouraged where possible (Plate 14).

In large feedlots it is costly to have very small groups but it is sometimes relatively easy to increase space allowances quite substantially, thus mitigating the worst effects of large group size.

Plate 14. Friesian bull beef unit divided into smaller pens

Trough space. Generally accepted allowances of trough space for various feeding systems are summarised in Table 7.7. It is obvious that where feed is given on a restricted basis then large space allowances have to be given so that all animals can stand side by side at the trough. It is also useful in such circumstances to have occasional partitions to prevent undue pressure when a long line of animals move.

Prevention of feed waste. This is largely a function of the design of the feeder, trough or rack. A barrier with slanted

Table 7.7. Guide to trough feeding space allowances for cattle

Situation	*Linear allowance per head (m)*
Self-feeding of silage to dairy cows	0.15–0.2
Self-feeding of complete diets to dairy cows, beef cows and large beef animals (400 kg+)	0.2
Self-feeding of silage or mixed diet to calves and young stock (up to 400 kg liveweight)	0.1–0.15
For restricted feeding of concentrates or forage at a trough or barrier. Cows and large beef animals (400 kg+)	0.6–0.7
For restricted feeding of concentrates or forage at trough or barrier. Calves and young cattle (up to 400 kg liveweight)	0.5–0.6

instead of upright divisions helps prevent the beast from lifting its head up and down and throwing food all over the place. Divisions of this kind also help to avoid bullying and pressure from aggressive neighbours.

Storage of feed at the feeding area.

Although it may be convenient to feed up to two or three times a day it is also sometimes useful to feed at less frequent intervals if storage space in the feeding area permits. High moisture diets such as grass silage and complete diets based on silage and other moist ingredients, can be stored for several days in northern European winter conditions. Under these conditions, where ambient temperatures are mainly around 5 °C or below, feeding frequency may be reduced to avoid weekend work or to ease feeding management. At the present state of knowledge the cattle feeder is wise to test this possibility with his own feed mixtures before risking spoilage of valuable feed.

REFERENCES

ANON (1978), 'Out of parlour cattle feeders', *Power Farming 57* (2), 23–31.

BATH, D. L. and BENNETT, L. F. (1980), 'Development of a dairy feeding model for maximising income above feed cost with access by remote computer terminals', *Journal of Dairy Science 63,* 1379–89.

BROWN, C. A. and CHANDLER, P. T. (1978), 'Incorporation of predictive milk yield and dry matter intake equations into a maximum-profit formulation program', *Journal of Dairy Science 61,* 1123–37.

COPPOCK, C. E. (1977), 'Feeding methods and grouping systems', *Journal of Dairy Science 60,* 1327–36.

FRIEND, T. H., POLAN, C. E. and MCGILLIARD, M. L. (1977), 'Free stall and feed bunk requirements relative to behaviour, production and individual feed intake in dairy cows', *Journal of Dairy Science 60* (1), 108–16.

FROBISH, R. A., HARSHBARGER, K. E. and OLVER, E. G. (1978), 'Automatic individual feeding of concentrates to dairy cattle', *Journal of Dairy Science 61* (12), 1789–92.

GREENHALGH, J. F. D. and REID, G. W. (1980), 'Complete diets for dairy cows: comparisons of feeding to appetite with rationing according to yield', *Journal of Agricultural Science 94* (2), 715–26.

GORDON, F. J. (1982), 'The effect of pattern of concentrate allocation on milk production from autumn calving heifers', *Animal Production 34,* 55–62.

GORDON, F. J., MCCAUGHEY, W. J. and MORRISON, B. (1981), *Milk Production,* Agricultural Research Institute of Northern Ireland.

HOLTER, J. B., URBAN, Jr. W. E., HAYES, H. H. and DAVIS, H. A. (1977), 'Utilisation of diet components fed blended or separately to lactating cows', *Journal of Dairy Science 60,* 1288–93.

JOHNSON, C. L. (1983), 'Influence of feeding pattern on the biological efficiency of high-yielding dairy cows', *Journal of Agricultural Science 100,* 191–200.

KROLL, O., OWEN, J. B. and WHITAKER, C. J. (1986), 'Variation in the complete diet composition given during the winter period to an autumn-calving herd', *Journal of Agricultural Science* 106, 297–306.

OWEN, J. B. (1979), *Complete Diets for Cattle and Sheep,* Farming Press.

OWENS, M. J., MULLER, L. D., ROOK, J. A. and LUDENS, F. C. (1978), 'Evaluation of magnetic grain feeder for lactating dairy cows', *Journal of Dairy Science 61,* 1590–7.

ØSTERGAARD, V. (1979), 'Strategies for concentrate feeding to attain optimum feeding level in high yielding dairy cows 482', *Beretning fra Statens Husdyrbrugs forsøg,* National Institute of Animal Science, Copenhagen.

SMITH, N. E. (1976), 'Maximising income over feed costs: evaluation of production response relationship', *Journal of Dairy Science 59,* 1193.

SMITH, M. E., UFFORD, G. R., COPPOCK, C. E. and MERRILL, W. G. (1978), 'One group versus two group system for lactating cows fed complete rations', *Journal of Dairy Science 61,* 1138–45.

TAYLOR, W. and LEAVER, J. D. (1982), 'A comparison of three methods of concentrate allocation to cows and heifers offered grass silage *ad libitum*', *Animal Production 34*, 359–60 (Abs).

WRAY, J. (1979), 'Feeding cows a complete diet', Bridgets Experimental Husbandry Farm *Annual Review No. 19,* 46–9.

Chapter 8

FEEDING PRACTICES—GRAZING

Grass utilisation presents a severe challenge to the cattle feeder and none more so than during the grazing season itself. Many of the problems associated with housing are eliminated but a new set of decisions has to be made. These can be avoided by adopting a conservative traditional approach to grazing practice but the need to make the utmost use of resources forces the feeder to wage a persistent battle against a complex and unpredictable set of circumstances.

CATTLE HAVE evolved as a grazing animal and over most of the world cattle systems involve a period of several months out on grass with little other feed. However, the length of the grazing season is highly variable, depending on climate and soil conditions, and it is common in certain areas, particularly under semi-arid conditions, to keep cattle on arable by-products and crops specially for cutting rather than grazing.

THE GRAZING SEASON

Under most temperate grazing conditions the season can usefully be divided into four periods each having their own particular problems for the cattle feeder.

1. The transition from winter feeding to grazing.
2. The early grazing period.
3. The mid-grazing period.
4. The late grazing period including the transition to winter feeding.

The Transition to Grazing

Cattle are normally turned out to grass in the spring when there is sufficient grass to sustain grass availability and when ground conditions are firm and dry enough to avoid poaching. Apart from climate and soil conditions the main factors that influence the early availability of pasture are the extent of previous grazing (particularly relevant in areas where sheep are used for the winter grazing of pastures), the use of fertiliser, especially nitrogen, and the species of herbage.

Previous grazing not only removes any winter growth that may be present but may retard spring growth if delayed beyond about 2–3 months before active spring growth commences.

Nitrogen fertiliser application can appreciably affect the early growth of pasture grasses and the timing of the fertiliser application is a key factor. Too early an application when soil temperatures are low and there is heavy subsequent rainfall can reduce the efficiency of use of the fertiliser. Workers in the Netherlands were the first to develop the 'T sum' calculation to enable a more precise assessment of the correct stage for N application. In this calculation the mean of the maximum and minimum daily temperature (if positive) is added each day from 1st January and N applied when the sum reaches a certain value e.g. T200.

Although the actual optimum value varies from area to area this value has been shown to be a useful guide to early fertiliser N use.

Species and variety within species can markedly affect availability of pasture in the spring. Where winter conditions are relatively mild, vigorous species like winter rye and Italian ryegrass are particularly early. In longer-term leys and permanent pasture the grasses are much earlier in growth than the clovers and varieties within species of grass can be chosen which can give appreciable differences in earliness.

Whatever the conditions applicable it is still an important question to decide when cattle should be turned out. In most cases early turnout can save winter feed, particularly relatively expensive concentrates. However this gain may be nullified if the overall picture is taken into account.

With store beef cattle it is sometimes the practice to turn them out before sufficient grass is available in the spring to allow them to 'grow with the grass' and to supplement them with hay or something similar in the early period. Being ruminants, cattle are sensitive to changes in diet since the whole rumen flora of micro-organisms has to be adapted to the new feed. One symptom of this is the profuse scouring that often occurs when cattle are suddenly turned out on to lush spring grass. Milking cows often suffer a minor fall in yield when first turned out, due to the disturbance in their habits, but in most cases milk yield shows a significant rise after a few days. This is usually accompanied by a change in milk quality, with a rise in SNF and falls in butterfat content. Increases in yield of 5–20 per cent are often experienced in dairy herds in the United Kingdom on turnout. The improved yield and SNF content are usually signs of an improved nutritional status and are particularly marked in cows that have been underfed either by having low-quality forage or low levels of concentrate feed in the late winter months. The fall in fat content is usually temporary and partly reflects a dilution phenomenon, since fat synthesis is slower to respond to improved feeding than the other components. It may also, in some cases, reflect the low fibre status of young spring grass, although this would not be expected to show up immediately.

Beef cattle on turn out can lose a substantial amount of body weight although this is largely a difference in gut contents. After a short period of adaptation cattle soon show a marked compensatory gain if winter feeding has been at a moderate level.

The most prudent practice to overcome the major problems of transition from indoor feeding to pasture is to employ a transitional feeding procedure whereby cattle are gradually accustomed over a period of 1–2 weeks to the new diet. This is done by continuing to feed the winter diet and allowing limited but increasing access to grass until full adaptation has taken place and the cattle are out day and night.

Other problems of the early transition period and beyond

include the possibility of disorders such as grass staggers (hypomagnesaemia) or bloat. These are discussed more fully in the next chapter.

The Early Grazing Period

Following the period of transition, cattle can then enjoy a period of pasture available in plenty and of a high quality. In most circumstances cattle can express near maximum performance levels at this stage. Spring-calved cows are capable of giving 40–50 kg of milk per day without supplementation and beef cattle of gaining 1–1.5 kg per day. Many pastures are capable of supporting in the region of one cattle unit (equivalent to a Holstein Friesian cow) on 0.2 ha of land, i.e. five cattle units per ha, at this stage (Plate 15).

Supplementation. Many experiments have been carried out to examine the use of supplementary feeding of cattle at this peak period of the grazing season (Holmes, 1980). Most of the results do not reveal any significant or economic benefit for such supplementation and although cattle will consume the concentrate, it simply replaces grass dry matter that they would otherwise have consumed (Table 8.1).

This is hardly surprising if the quality of the grass is taken into account. The more perceptive workers in this field have examined supplementation in relation to stocking rate and this aspect is discussed more fully later in the chapter.

An extreme form of supplementation allied to high stocking rate is the 'buffer' feeding of dairy cows developed at the School of Agriculture, Aberdeen in the 1970s. In this method pasture stocking rates are raised from 5 per hectare to 10–15 and access is given to an ad libitum supply of a complete diet formulated to be less palatable than grass, so that only a relatively small quantity is consumed when grass is available in quantity. Table 8.2 summarises some of the results.

The Mid-season Period

After 1½–2 months of the grazing season some deterioration usually sets in, particularly in pasture quality. This is the result of several factors including soiling by cow dung, uneven

Table 8.1. Summary of experiments in which cows at grass were offered supplementary concentrated feeds

Reference	*Duration of trial*	*Supplementary feed*	*Daily amount of feed*	*lb feed/ 1 lb extra milk**
MacLusky (1955)	Short-term	Balanced conc.	8 lb/cow	2.7
Wallace (1957) Expt. 2	8 wk	Balanced conc.	6 lb/cow	2.2
Holmes (1958)	May–Sept.	Balanced conc.	8 lb/cow	4.0
Corbett (1958)	Short-term latin square	Flaked maize	8 lb/cow	3.0
Seath *et al.* (1959)	18 wk	Grain meal	5.5 lb/cow	2.5
	18 wk	Grain meal	11 lb/cow	2.5
Castle *et al.* (1960)	July–Oct.	Oats and flaked maize	6 lb/cow	3.3
Laird & Walker-Love (1962)	June–Aug.	Balanced conc.	4 lb/gal over 4—3 gal	1.0
	May–Oct.	Balanced conc.	(Reduction as season progressed)	2.2
Shepherd (1962)	May and June	Barley	7 lb/cow	4.1
		Balanced conc.	7 lb/cow	2.4
Castle *et al.* (1964)	May–Oct.	Balanced conc.	According to milk yield, on average 2.1 lb/gal	1.8
Bryant *et al.* (1965)	22 wk in 1962	Balanced conc.	1 lb/6.8 lb FCM	1.2
	16 wk in 1963	Balanced conc.	1 lb/3.2 lb FCM	1.6
Wood (1966)	July–Sept.	Balanced conc.	3.75 lb/gal over 3 gal	6.7
			3.75 lb/gal over 1.5 gal	5.7
Hutton & Parker (1967)	Short-term changeover trials	Balanced conc.	2.2 lb/cow	1.8
		2 cows/acre	4.4 lb/cow	2.1
		Balanced conc.	2.2 lb/cow	2.2
		2.6 cows/acre	4.4 lb/cow	1.6
Frens & Bosch (1950)	June–Oct.	Sugar beet pulp	8.8 lb/cow	5.1
Sjollema (1950)	Survey of 14 experiments	Sugar beet pulp	6.6–11.0 lb/cow	2.9
Hart (1956)	Survey of 11 experiments	Sugar beet pulp	2.2–6.6 lb/cow	4.0
Corbett (1958)	Short-term latin square	Sugar beet pulp	8 lb/cow	3.6
Corbett & Boyne (1958)	May–July	Sugar beet pulp	8 lb/cow	6.1
	Aug.–Sept.	Sugar beet pulp	10 lb/cow	3.4

* Assuming that the difference between the mean milk yields of the cows receiving and not receiving supplements was completely attributable to the supplementary feed.

Source: Leaver, J. D., Campling, R. C. and Holmes, W. (1968), *Dairy Science Abstracts 30,* 355–61.

grazing resulting in the accumulation of mature and senescent plant material. There is also a poorly understood additional

Plate 15. (*above*) Productive cows on summer grass (*Fisons Ltd.*)

Plate 16. (*below*) Early bite being strip grazed in Cornwall (*Farmers Weekly*)

Table 8.2. Buffer feeding of dairy cows at grass

	Control group	*Buffer group*
No. of cows	30	30
Cows/ha grass	4.9	7.4
Concentrates/cow (kg)	320	815
Mean daily milk yield (kg)	18.2	17.7

Source: North of Scotland College of Agriculture (1976), *Research, Investigations and Field Trials 1974–75.*

seasonal factor which results in some depression in animal DM intake and in the efficiency of conversion of ME. This is not wholly reflected in the reduced digestibility of the pasture material but changes in some of the micro nutrients and in the soluble carbohydrate fraction may be associated with the change.

For most beef animals and for autumn-calved cows, in late lactation, adjustment of stocking rate to ensure a sufficient supply of pasture still maintains intake and performance beyond the point at which supplementation is profitable. However spring-calved cows may benefit from supplementation and it is possible that an adequate UDP supply may be important to maintain lactation and to minimise the steep fall in the lactation curve that often occurs at this point.

At the present state of knowledge, judging the rate of supplementation is more of an art than a science but it is becoming increasingly evident that much of the reduction in milk production of spring-calving cows stems from nutritional neglect from mid lactation onwards.

The Late Grazing Period

This is a period when the problems of pasture quality just noted are accentuated and climatic conditions and ground conditions become increasingly more difficult. Decision-making at this time is very complicated since there is a lack of objective measures of grass quality and availability. The decisions to be made depend very much on the system, for instance the calving period of a dairy herd. In autumn-calving herds calving usually coincides with the stage at which the cows are making the transition to indoor feeding and this in some ways eases the problem. However certain major disorders such as milk fever, hypocalcaemia and laminitis may be associated with factors operating in the autumn-calving herd at this time, and are discussed in the next chapter.

Ideally, the autumn-calving cow should be kept in at night from the time that she calves or shortly before that.

Cows in spring-calving herds present quite a delicate problem and there has been a tendency to assume that their nutritional needs are not critical since they are in late lactation and therefore according to traditional theory have low 'requirements'. This philosophy undoubtedly is largely to blame for the rapid falls in the lactation curve and short lactations of spring-calving cows. There is now evidence to show that cows in late lactation respond profitably to better feeding.

As with the spring transition period, it is common and wise to bring cattle in gradually, allowing them access to grass during the day for some time after they are housed at night.

With beef cattle similar difficulties are experienced with judging the right time to start winter feeding and to bring cattle in. The decision is partly influenced by the type of animal and the marketing objective. Cattle in good body condition that are to be marketed early in the winter period would need to be watched carefully to avoid any loss of condition before they are brought in. On the other hand younger store cattle that are to be grazed after the winter period, or heifers not near to calving, can be left out rather longer and wintered at a lower level of feeding.

Grazing Systems

Several grazing systems are potentially available for cattle. They are basically of two types.

1. Rotational

This is where the grazing area is subdivided into smaller areas and the cattle moved from one small area to another in a cyclical fashion so that when one circuit is completed the cattle go on to the first area vacated some time earlier.

This system is subdivided into several varieties.

(a) field-by-field rotation—where the subareas are rather large and may number anything from three to eight in total. Cattle stay in each field for several days at a time.

(b) paddock system—this usually involves the subdivision of fields into smaller units and the total area might involve 12–20 paddock units. Cattle stay on each paddock for one or perhaps two days.

(c) strip grazing—this system is usually superimposed on systems (a) and (b) whereby the grazed units are made even smaller by moving an electrified fence often twice a day (Plate 16). If the fields within which it is operated are large, a back fence may be employed to prevent cattle grazing the regrowth that shows itself on strips a few days after they have been grazed.

All the rotational systems may be modified to allow preferential access to high-yielding cows as opposed to low-yielding or dry cows. This is simply achieved by giving the preferred group the first stay in an area moving them on to allow the second group to consume the pasture, such as a leader–follower system.

Continuous Grazing or Set Stocking

This system involves giving the animals continuous access to all the grazing area allotted to them. The same area as is being grazed rotationally in the rotational grazing system is thrown open, either by having no fences or leaving all the gates open.

Under the rotational system the pasture is being subjected to a series of defoliations interspersed with a protected period for recovery. Under the set-stocking system, cattle themselves operate as a regular rotational mowing gang covering every square metre of land regularly. The result of this treatment is that the pasture plants adopt a prostrate habit with much of the growth at a horizontal rather than a vertical angle, giving a thick, highly tillered sward. This sward is always well covered with green leaves as compared with the pale stemmy stubble left after the intense periodic grazing on the rotational system.

Choosing a Grazing System

In choosing between the various grazing systems available there are many factors to consider:

1. Yield of animal products. Although experimentation is difficult and expensive several useful experiments have been carried out to examine the relative yield of milk or of beef from the use of the various systems.

The general consensus, taking all the evidence into account, is that there is little consistent advantage to adopting rotational as opposed to continuous grazing, other things being equal.

2. Suitability for Fertiliser Application. In spite of early fears that high levels of fertiliser application were not consistent with continuous access to grazing, there is now enough experience to confirm that fertiliser application poses no problem. Provided the fertiliser is applied in small regular applications, say every three to four weeks during the growing season, no special precautions have to be taken to prevent consumption by the cattle. It is a different matter with larger application of materials like basic slag where it is important to confine the application to the non-grazing period.

3. Flexibility for conservation. Since the rate of grass growth is highly variable in the season it is important, in order to achieve high efficiency of grass use, to be able to vary the area available to a given number of cattle. In the rotational system this can be done by taking some of the fields or paddocks out of the rotation and returning them after conservation.

Since continuous grazing or set stocking is usually based on existing fields, with the gates open, exactly the same thing can usually be done with part of the area being out of commission whilst undergoing conservation and the remainder of the area remaining under continuous grazing.

4. Capital cost and management ease. Rotational systems may involve extra expense in fencing and provision of watering points but the main attraction to the advocates of both systems are aspects of ease of management. Rotational enthusiasts claim that they know better what they are doing whilst set-stocking practitioners claim that they can make smoother adjustments in feed availability amongst other things.

The main aspects in favour of continuous grazing, which accounts for the fact that it has only been briefly threatened by rotational systems, is that as the traditional simple system, is well suited to the natural herd behaviour of cattle, the alternative has offered no consistent tangible benefits and can bring several new problems in its train.

The Utilisation of Pasture

The ultimate aim of the grazier is to make the most efficient use of his resources and thereby maximise overall farm profit. A major consideration is to decide on the optimum combination of fertiliser application, stocking rate and feed supplementation so that the grass is not only consumed completely without waste but that it is utilised profitably.

A useful way of considering this problem is that in Fig. 8.1.

It simply shows that at low stocking rates, whilst production of the individual is at its highest in the short term, production per hectare is low, grass is not fully utilised, leading to deterioration and loss. At very high stocking rates on the other hand, the pasture is fully utilised, perhaps even suffers excessive hoof damage and fouling, but since it is overstocked cattle do not perform well and overall performance per hectare is again low.

It is clear in this simple picture that the aim of each grazier is to achieve the magic point at which the pasture is fully consumed, i.e. pasture reserves or herbage mass are not increasing but where all cattle are able to consume their fill. At such a point both individual production and production per hectare are at their highest.

The picture itself is too simple for practice because it is quite impossible for any grazier to manage his pasture so that

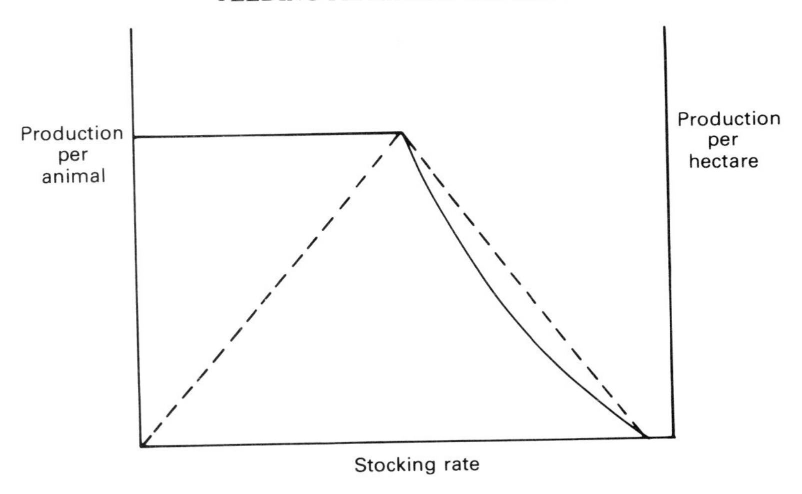

Figure 8.1. Theoretical relation of production per animal and per hectare to stocking rate

he is consistently at the magic critical point. However the diagram sets a clear objective in pasture management and it sets a logical framework for considering the effects of fertiliser application and feed supplementation.

In this context the main cornerstones of efficient pasture management can be summarised as follows:

1. Pasture Availability

For any class of stock a target level of pasture availability should be set and, within reason, this should be maintained throughout the grazing season. Where, for reasons beyond control (for instance at the beginning and end of the season or during periods of drought), this availability cannot be maintained, then supplementation should be carefully considered.

There is no published evidence to suggest that there is any important difference between the desired target pasture availability of different classes of cattle such as dairy cattle, beef cattle or calves. It is assumed therefore that one value can be used for all. The main evidence on the critical level of pasture availability comes from work on dairy cattle.

Figure 8.2 indicates the conclusion that optimum pasture length for cows should be in the region 7–9 cm.

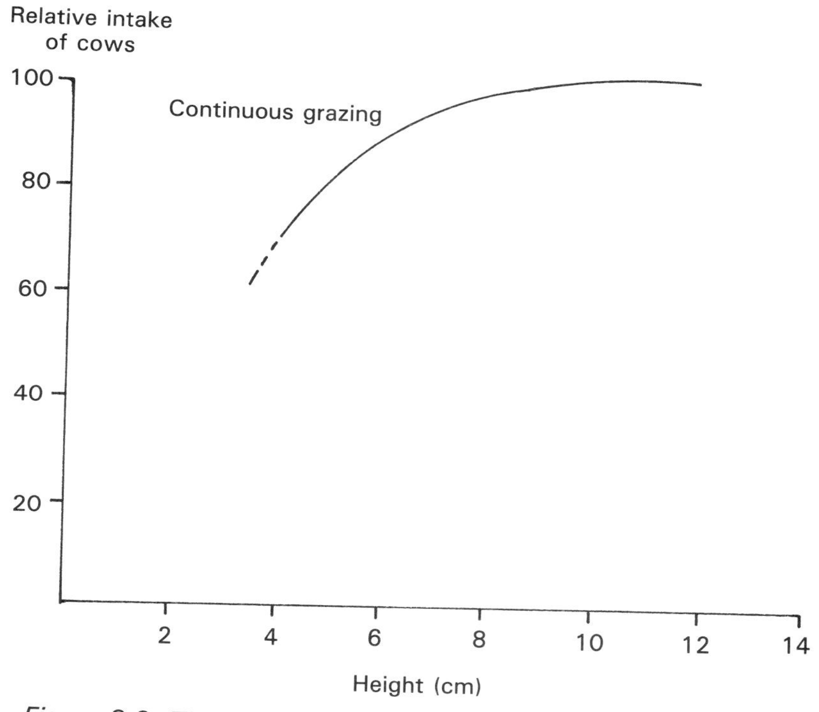

Figure 8.2. The relative intake of cows in relation to sward height (extended tiller height) on set-stocked swards
Source: R. D. Baker (1986)

2. Rate and pattern of fertiliser application
Assuming that pasture soil values for pH, potash and phosphate are kept within the ranges normally associated with optimum production, much of the variation in pasture growth is influenced by nitrogen application. Unfortunately very little evidence is available on the effects of N fertiliser application other than from studies of the yield of cut herbage. These are by no means to be taken as reliable guides to the effect on animal production.

(a) Annual application rates. Fig. 8.3 shows the effect of N fertiliser rates on the DM production of grass and grass/clover swards. Although these experiments indicate an almost consistent linear response up to application rates of about 400 kg/ha the little evidence available from animal production responses would indicate a lower optimum for both dairy cattle production and for beef animals.

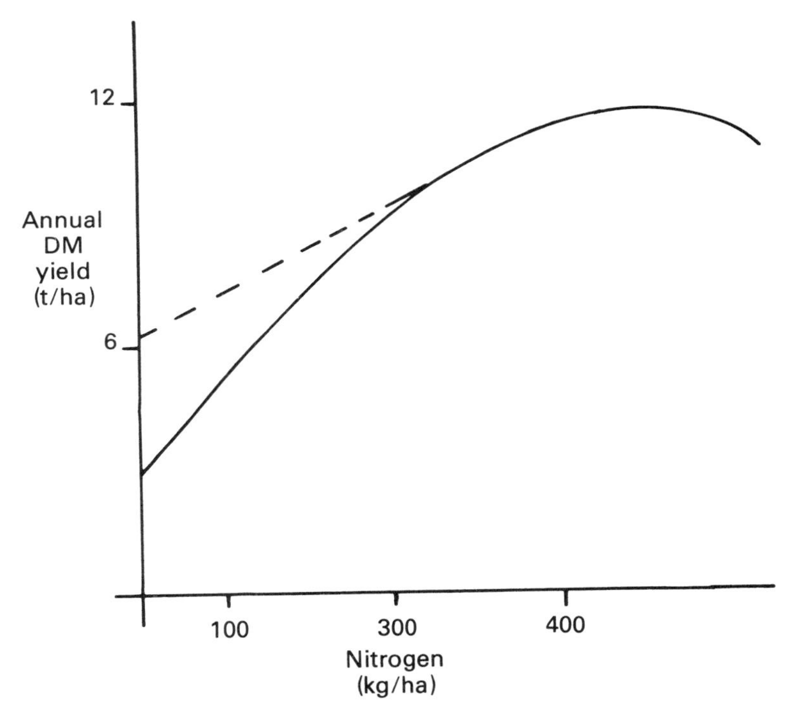

Figure 8.3. Response of grass and grass/clover swards to N fertiliser input

(b) Pattern of application. Within limits the seasonal pattern of pasture growth can be manipulated by the fertiliser application pattern as shown in Fig. 8.4.

This can be overdone and a loss of efficiency of N use can result. It is important therefore to consider N application

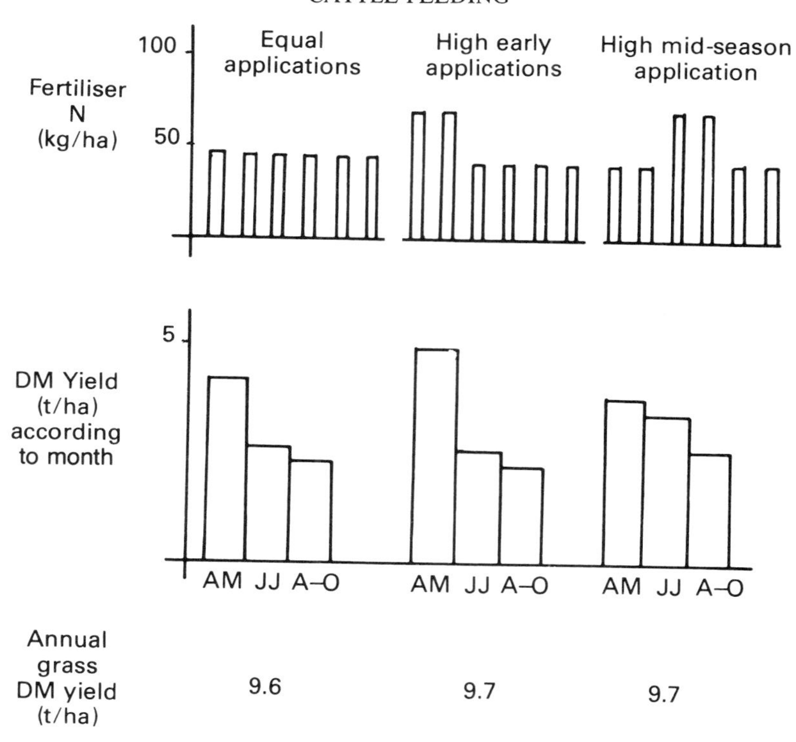

Figure 8.4. Pattern of grass growth in relation to distribution of fertiliser N application

Source: Thomas & Young, 1982, ICI Agricultural Division, GRI Hurley

pattern in the context of the overall supply of grass taking conserved material into account.

3. Conservation

A major tool in pasture management is the use of conservation to lop off the peak of grass growth in the early growing season for use in the winter months. This allows a contraction of the area available for grazing when pasture growth is at its peak and a subsequent expansion to coincide with the tailing off of growth later in the season. This tool can be highly flexible since conservation may be taken in several cuts during the season.

Plate 17. (*above*) Estimating grass availability for management decisions (J. B. Thomas, UCNW)

Plate 18. (*below*) Strip grazing kale with a dairy herd in South Wales

Table 8.3 shows the area required for grazing and for conservation in various cattle production systems in the United Kingdom.

Without conservation pasture management would be a much more difficult task.

Table 8.3. Average grazing and conservation areas (ha) required for various livestock systems under U.K. conditions

	Dairy cows (Friesian)	*Suckler beef cows*	
		Spring calving	*Autumn calving*
Grazing area per animal	0.31	0.33	0.36
Conservation per animal	0.26	0.27	0.29
Total forage area per animal	0.57	0.6	0.65

Source: Nix, J. (1982), *Farm Management Pocketbook,* Wye College.

4. Supplementation

Supplementation at grass needs to be carried out with care. Looking at the diagram in Fig. 8.1, it is evident that, if the system is not at the critical point in terms of stocking rate, then supplementation is wasted unless the quality of the grass is in some way deficient. However it is also apparent that supplementation may be a very useful fine-tuning tool to avoid unwanted checks to animal production when stocking rate inadvertently exceeds the critical point. One way of supplementing or replacing pasture in the autumn is to grow special crops such as kale (Plate 18).

Measurement of Pasture Utilisation Efficiency

When using the principles outlined in the foregoing section it is a useful exercise to be able to record the system accurately. This is not an easy task but a recent useful innovation is the system of assessing utilised ME (UME) and pasture utilisation efficiency. The basic data to be recorded are the areas under consideration as divided into various fields and the use made of each field in terms of grazing and conservation. Each area is therefore assessed in relation to the recorded animal production from it in terms of grazing and conserved material. From this, together with the theoretical ME requirement of the classes of stock the total UME is calculated. The theoretical target production for the land in relation to its potential is then estimated and the UME expressed as a fraction of the target ME supply, to provide the utilisation efficiency figure.

For example a spring-calving dairy herd, giving an average yield of 5000 kg milk per cow at a stocking rate of 2.2 cows per hectare (grazing + conservation) and a concentrate input of 1200 kg per cow.

Estimated ME requirement per cow is 25 GJ for maintenance and pregnancy and $0.0052 \times$ milk yield i.e.

$$25 + 0.0052\ (5000) = 51 \text{ GJ}$$

Therefore the UME from grass is $51 - (1200 \times 0.012) = 36.6$ per cow (assuming that the energy value of the concentrates is 0.012 GJ per kg).

$$\text{UME per ha} = 36.6 \times 2.2 = 80.52 \text{ GJ}$$

Target UME value for medium textured soils in high rainfall areas (400 mm+) at high levels of N fertiliser use is in the region of 100 GJ per ha.

Thus estimated utilisation efficiency is

$$\frac{80.52}{100} \times 100 \simeq 80\%$$

Practical difficulties in implementing this procedure arise from the difficulty of estimating forage area attributable to an enterprise on a mixed farm. Also if values are used for comparison between farms and enterprises the accuracy of the assumptions become crucial.

References

ALLEN, D. and KILKENNY, B. (1980), *Planned Beef Production*, Granada Publishing Co.

BAKER, R. D. (1986), 'Advances in cow grazing systems', *Grazing* (edited by J. Frame), *British Grassland Society Occasional Symposium No. 19*, page 160.

CARTER, W. R. B. (1960), 'A review of nutrient losses and efficiency of conserving herbage on silage, barndried hay and field cured hay', *Journal of the British Grassland Society 15*, 220–30.

CASTLE, M. E. and WATKINS, P. (1979), *Modern Milk Production*, Faber and Faber.

GREENHALGH, J. F. D. (1975), 'Factors limiting animal production from grazed pasture', *Journal of the British Grassland Society 30*, 153–60.

GREENHALGH, J. F. D., REID, G. W., AITKEN, J. N. and FLORENCE, E. (1966), 'The effects of grazing intensity on herbage consumption and animal production. 1. Short-term effects in strip-grazed dairy cows', *Journal of Agricultural Science 67,* 13–23.

GREENHALGH, J. F. D., REID, G. W. and AITKEN, J. N. (1967), 'The effects of grazing intensity on herbage consumption and animal production II. Longer-term effect in strip grazed dairy cows', *Journal of Agricultural Science 69,* 217–23.

HODGSON, J. and JACKSON, D. K. (1975), 'Pasture utilisation by the grazing animal', *Occasional Symposium No. 8,* British Grassland Society.

HOLMES, W. (1980) Editor, *Grass—its Production and Utilisation,* Blackwell Scientific Publications.

HORTON, G. M. J. and HOLMES, W. (1974), 'The effect of nitrogen, stocking rate and grazing method on the output of pasture grazed by beef cattle', *Journal of the British Grassland Society 29,* 93–9.

LAIRD, R., LEAVER, J. E., MOISEY, F. R. and CASTLE, M. E. (1981), 'The effect of concentrate supplements on the performance of dairy cows offered grass silage ad libitum', *Animal Production 33,* 199–209.

LEAVER, J. D., CAMPLING, R. C. and HOLMES, W. (1968), 'Use of supplementary feeds for grazing dairy cattle', *Dairy Science Abstracts 30,* 355–61.

MCDONALD, I. W. (1968), 'The nutrition of grazing ruminants', *Nutrition Abstracts and Reviews 38*, 381–400.

M.L.C. (1978), 'Grazing Management', *Beef Production Handbook No. 4,* Meat and Livestock Commission.

OWEN, J. B. and RIDGMAN, W. J. (1968), 'The design and interpretation of experiments to study animal production from grazed pasture', *Journal of Agricultural Science 71,* 327–35.

THOMAS, C. and YOUNG, J. W. O. (1982), *Milk from Grass,* ICI Agricultural Division and G.R.I.

WESTON, R. H. and HOGAN, J. P. (1975), 'Nutrition of Herbage-fed ruminants', *The Pastoral Industries of Australia,* edited by G. Alexander and O. B. Williams, Sydney University Press.

WHEELER, J. L. (1962), 'Experimentation in grazing management', *Herbage Abstracts 32,* 1–7.

WILKINSON, J. M. and TAYLOR, J. C. (1973), *Beef Production from Grassland,* Butterworths.

Chapter 9

FEED-RELATED HEALTH PROBLEMS

Cattle, particularly the modern dairy cow, are subject to pressures on three main fronts. First there is the cow's own nature, bred for generations towards higher production. This then is increasingly exploited by management and feeding practices that push the inherent tendency to its limit. Finally all this is often carried out under conditions which can severely test the cow's constitution. Housing conditions are designed to meet several criteria and there is room to fear that some modern systems do not please the user cow.

SEVERAL MODERN problems of ill-health are complex and have many causes. Some are not affected by feeding but a few are clear consequences of malnutrition or undernutrition. In between these two extremes is a grey area where feeding is only one of the known or probable causes of a disorder. Even when the basic nature of the disorder is thought to be well understood there may not be a clear answer in the form of either prevention or cure. When a disorder is poorly understood and nutrition is only a suspected contributory factor, it is natural that firm guidelines do not exist.

The purpose of this chapter is to look further at those health problems where feeding practice is clearly implicated and to outline some of the principles of avoiding less specific nutritionally influenced problems.

Milk Fever (Hypocalcaemia)

This disorder belongs to those that are termed metabolic disorders or sometimes production disorders. Like many of the others it is complex in its causation and is far from being a simple deficiency disease.

The traumatic manifestation of the disorder in the newly calved cow ('downer' cow) is associated with a fall in blood calcium levels to below 2.0 and usually below 1.5 m moles/l serum. It is often also associated with a fall in the serum inorganic phosphate level and sometimes with a hypomagnesaemia i.e. a lowered serum magnesium level.

Its cause is the sharp rise in the demand for calcium set by the sudden onset of milk secretion and the failure of the cow's metabolism to cope with this sudden extra demand. The demand is normally met by the dietary intake of calcium with absorption from the intestine varying according to requirement. Fluctuations above this level are taken care of by the mobilisation of calcium from body reserves, of which 90 per cent is contained in the skeleton. These systems are chiefly influenced by vitamin D, parathyroid hormone and calcitonin. Under normal circumstances these mechanisms ensure normal body operation, even when demands for calcium fluctuates within quite wide limits—mostly as a result of milk secretion.

It is in the older cow that the mechanism usually breaks down, presumably due to the higher milk secretion rate and the impaired ability to mobilise calcium from the more mature skeleton.

Treatment is now reasonably standardised to the slow intravenous administration of 400 ml of 40 per cent calcium borogluconate solution in the case of simple milk fever or with the addition of an intravenous injection of 400 ml 5 per cent calcium hypophosphate and subcutaneous injection of 400 ml 25 per cent magnesium sulphate when these complications are suspected. Such treatment is usually and dramatically effective, although in some cases the cow may suffer one or more relapses.

Prevention is far more debatable and there is some controversy as to what is the best method to adopt. It is now considered that a low-calcium diet in late pregnancy (less than 0.5 per cent of calcium is in the diet dry matter) which is changed to a high-calcium diet (1 per cent) two days before calving, is a helpful regime. The low-calcium diet may keep the calcium-conserving systems functioning whilst demand is low and the high level of dietary calcium provided as milk secretion starts helps to reduce the deficit involved.

The execution of such a system in practice can be difficult, particularly when the cow is at pasture during late pregnancy, for grass usually contains more than 0.5 per cent calcium.

It is also possible that milk fever is exacerbated by fat body condition at calving brought about by 'steaming up' since this depresses feed intake whilst at the same time involving high initial milk secretion.

The use of injections of vitamin D_3 in the period between eight and two days before calving has been shown to affect the incidence of milk fever but the substance is toxic and the timing of the injections is difficult in practice. The use of a synthetic analogue of vitamin D_2 (1 cholecalciferol) shows more promise of being suitable in practice but is not yet widely used for the purpose.

Grass Staggers (Hypomagnaesemia)

This is another disorder of a similar kind to milk fever and, as mentioned above, these two disorders can sometimes occur simultaneously. Grass staggers is associated with low levels of blood magnesium and is again mainly a disorder of the lactating cow although it can be brought about in calves on a diet very low in magnesium. Although it can occur at calving, often in association with hypocalcaemia, it more often occurs some weeks later. Its manifestation can vary from the acute form when the cow may be found dead or in a tetanic spasm or it may take a more chronic form, associated with depressed feed intake which is difficult to diagnose.

Its cause is an imbalance between the demand for magnesium, heightened by milk secretion and the supply of magnesium available to satisfy the demand. There is little effective storage of magnesium since the magnesium deposited in the skeleton is unavailable as a reserve except in young calves. A steady daily supply of magnesium in the diet is therefore essential. The precise combination of factors that precipitate hypomagnesaemia are difficult to discern with clarity.

A simple deficiency of magnesium in the diet can cause

hypomagnesaemia but more often dietary magnesium level is confounded by variation in the availability of the magnesium. A high-yielding cow on spring pasture is subject to a large demand for magnesium at a time when dietary levels on the grass may be only moderate but more significantly when absorption from the gut is depressed (Fig. 9.1).

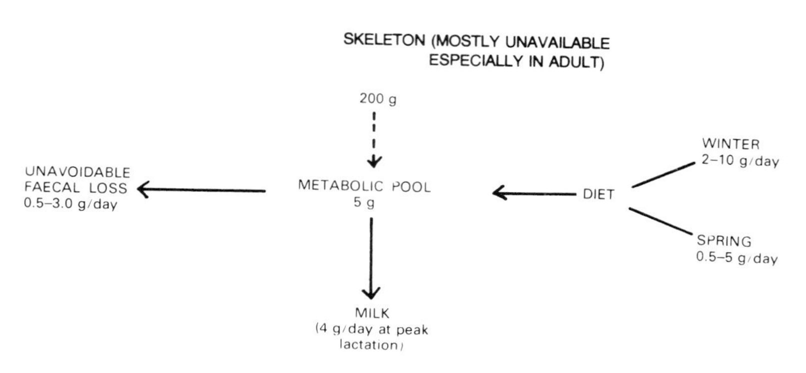

Figure 9.1. Magnesium in the cow's body
Source: Bowers, H. B. (1968) PhD thesis, University of Aberdeen

Hypomagnesaemia may also occur, particularly in beef cows suckling calves, on autumn and winter pasture.

It is now known from work initially carried out in the Netherlands that potassium and nitrogen levels in the soil interact to depress the level and availability of herbage magnesium. Thus application of potash fertilisers or of slurry to cow pastures in the spring may precipitate hypomagnesaemia in the grazing cows.

Hypomagnesaemia can sometimes develop, particularly in the case of nursing beef cows, when undernutrition is present and it can often be precipitated by stresses such as a sudden cold snap.

Prevention. The main preventative measures advocated are the daily administration of magnesium, preferably in the form of a high magnesium concentrate supplement, accompanied by a generally adequate feeding level (in terms of energy).

Magnesium is often incorporated in the form of calcined magnesite to ensure a consumption of about 50 g per cow per day.

Other measures have their exponents but are more debatable in their value. The administration of magnesium to the pasture (e.g. 32 kg calcined magnesite per ha) is one, but frequency and quantity required are not clearly established. The use of magnesium bullets which are intended to lodge in the rumen and to release magnesium gradually over a period of time is another. Whilst fine in theory they may be regurgitated or dislodged and the quantity of magnesium in the bullets is small.

Administration of magnesium salts in the drinking water also has its advocates but the consumption of water from the intended troughs on lush spring grass is unpredictable.

Ketosis

This disorder has more than one consequence and occurs in several guises. Pregnancy toxaemia, more commonly associated with sheep, is one form, primary bovine ketosis and 'fatty liver' or 'fat cow' syndrome, others. The name comes from the accumulation of ketone bodies in the cow's system which can be recognised from the sweet odour of acetone on the cow's breath. This is a symptom that shows that the animal's energy metabolism, particularly that associated with the utilisation of body fat during periods of high energy demand, is not functioning properly. This is a serious position for the cow since the supply of blood glucose can be impaired through the decrease in the production of propionic acid from the rumen and supply of amino acids from the intestine. A severe upset in the blood glucose supply can eventually affect the brain and lead to the classical nervous symptoms often associated with this disorder.

Primary bovine ketosis is the most commonly recognised form. It usually occurs between the second and the seventh week post-calving when the cow is approaching peak production, at a time of substantial body weight loss when the cow's system becomes unable to cope with the demands on it. A certain degree of weight loss is normal but if this becomes too severe, or there is some upset that depresses food intake, the delicate balance between energy supply and demand breaks down and a critical vicious circle sets in. Intake of food and water is abruptly or quickly (2–4 days) suspended and this, by cutting off milk secretion, allows the system a chance of natural recovery. If the attack is severe then the nervous signs may manifest themselves indicating impairment of brain function.

Conditions conducive to the development of ketosis include overfeeding in late lactation and the dry period so that cows are in fat condition before calving. Such conditions can also lead to the development of 'Fatty liver' or fat cow syndrome when excessive mobilisation of body fat reserves leads to the transport of non-esterified fatty acids to the liver. The liver of cows with severe symptoms may have up to 70 per cent of fat with a consequent severe effect on liver function and a virtual cessation of gluconeogenesis which is the only way the fat could be metabolised.

Prevention. The prevention of this upset to energy metabolism involves several measures:

1. Adequate pregnancy feeding whilst avoiding overfatness in late pregnancy, to achieve a fit cow in condition score $2\frac{1}{2}$–$3\frac{1}{2}$ at calving.
2. Supply of an adequate diet *ad libitum* with particular attention to UDP levels.
3. Avoidance of excessive concentrate intake that can depress roughage intake below critical levels. Such feeding can lead to subclinical acidosis, fluctuating appetite, and misplaced abomasum, all ideal precipitating factors for ketosis.
4. Avoidance of sharp changes and stresses at critical periods especially in the first two months of lactation.

Acidosis

Overfeeding of concentrates, both sudden or protracted, can lead to a disorder called acidosis. As its name suggests it is associated with the production of very high levels of lactic acid in the rumen; there may be several other critical accompanying factors. The most acute form is that experienced when the animal accidentally gorges itself in the feed bin and this usually results in death within a few hours of the occurrence.

A frequent circumstance when the acute form develops is when a concentrate supplement is introduced to a previously unsupplemented group of cattle. When a group of cattle are introduced to a grain supplement and only a few eat it then those few may eat too much, in spite of the fact that only a small allowance per animal was given for the whole group of cattle.

The less acute form of acidosis occurs in dairy cows or beef cattle that are given a high-concentrate diet; some cattle under such circumstances may eat too little forage. Such mild acidosis may have few overt symptoms unless a careful record of feed intake is kept. Cattle on all-forage diets normally have a rumen pH of about 6.5 and a rumen flora that maintain a rumen fluid with VFAs approximately 70% acetic, 20% propionic, 8% butyric and 2% others, but with very low levels of lactic acid.

Increased concentrate levels enhance the supply of starch or readily fermented carbohydrates. This causes the normal gram negative rumen bacteria to be replaced by gram positive bacteria such as lactobacilli which ferment the starch to lactic acid resulting in low pH. Such conditions lead to rumenitis which damages the rumen wall leading to an absorption of toxic substances into the system.

Several consequences flow from such a marked abnormality of function. Feed intake may show increased fluctuation, laminitis may be induced (as described below), and the low milk fat syndrome in dairy cows is a consequence. Overall efficiency of lactation suffers and even in beef animals the rumenitis can lead to several unwelcome consequences such as liver damage (hepatitis).

Treatment of acute cases is difficult. Large oral doses of water, possibly containing antacids such as magnesium carbonate, administration of antibiotics by mouth and injections and infusions of saline fluids to overcome dehydration may be necessary. Since deactivation of thiamine by thiaminase is one complication, administration of the B complex vitamins is normally advised.

Prevention. This is clearly related to observing a cautious forage:concentrate ratio and if tables such as those in Chapter 5 are consulted it seems unlikely that, where good forages are used, the forage proportion of the diet DM should fall below 40 per cent for maximum profitability. Where a low forage diet is contemplated the possible deleterious effects should be carefully considered.

Measures such as mixing the diet, as in a complete diet system and frequent feeding, where high concentrate levels are employed, are also likely to safeguard certain cattle from the danger of high variation in the intake by individuals.

Some work supports the use of antacids such as sodium bicarbonate in the diet at 1–2 per cent of the DM in circumstances when high proportions of concentrates in the diet are necessary. However it appears likely that diets formulated for the most economic conversion of diet into products are unlikely to warrant such a practice.

Laminitis

This is an inflammation of the quick of the hoof or the horn growing layer (corium) which is one of the chief contributors to chronic lameness in cattle. It is a painful condition for the animal and its consequences in hoof malformation and disturbed gait all contribute to a substantial loss, particularly for the dairy farmer.

Several factors contribute to the damage involved in laminitis. These include metabolic abnormalities, largely brought about by incorrect feeding, the traumatic factors such as external bruising from rough concrete, continual exposure of the hoof to liquid slurry and toxic effects such as those following mastitis or uterine infections. All these factors combine with the genetic susceptibility of the individual to cause laminitis. Any factor such as uncomfortable lying quarters, which increase the proportion of the day which the cow stands in slurry, is likely to exacerbate the condition.

A dietary cause for some laminitis has long been suspected and excessive concentrate consumption or high protein intake have been amongst the factors most suspected. Because of the difficulty of experimentation in this field hard evidence is difficult to obtain but Livesey at the University

College of North Wales, Bangor farm, has demonstrated clearly that laminitis can be induced by increasing the concentrate:forage ratio as shown in Table 9.1.

Table 9.1. The effect of lactation feeding on the incidence of laminitis in dairy cows given complete diets

	50:50 concentrate to forage ratio	*60:40 concentrate to forage ratio*
Number of cows	26	25
Number and percentage of cows showing laminitis	2 (7%)	17 (68%)

Source: Livesey, C. T. and Fleming, F. L. (1982), Personal Communication.

This reinforces the need to look carefully at any cases where high-concentrate or high-protein diets are employed and particularly to avoid rapid changes in diets in this respect. Much more work is required in this area before the farmer can be assured of firm guidelines for the prevention of lameness.

Bloat

Bloat is another metabolic disorder of cattle that is feared because of its acuteness and the severe problem it can create in localised situations. It is caused by the formation of a stable foam in the rumen which cannot be released by the normal process of belching (eructation) resulting in a rapid build-up of pressure and death from circulatory and respiratory breakdown. It is another of the complex conditions which were much underestimated in the early stages of their study but have proved to be difficult to understand fully and therefore to deal with. Bloat is much associated with the grazing of pastures containing legumes such as clover and lucerne (alfalfa) and it is no wonder that it has been of major economic concern and a subject of much study in New Zealand where surveys have shown that 48–80 per cent of dairy herds suffer annually (Table 9.2).

As with other metabolic disorders, animal factors play their part so that some animals when subjected to certain

Table 9.2. Incidence of bloat in New Zealand dairy herds

Herds affected (%)	80
Total no. of herds	8,599
Mortality (%)	1.2

Source: New Zealand Dairy Board, 1965.

conditions are more susceptible than others.

Of the plant factors it is known that the main species involved are the clovers *(Trifolium repens* and *pratense)* lucerne *(Medicago sativa)* and the natural 'medic' legumes (*Medicago* spp.). However other species such as grasses can sometimes be involved and occurrences have been observed with indoor feeding. The incidence of bloat from New Zealand experience has been shown to be highly erratic and although many outbreaks are associated with young, immature herbage several other factors such as the climate may well be involved in a way not yet clearly understood.

When the conditions are conducive to bloat the animal eats the material and the rumen microbes ferment the material at a rate and in a form which cannot be accommodated. Progress in the development of methods of treatment and prevention has been slow but several practical measures can now be taken to avoid its worst consequences.

Methods of Preventing and Controlling Bloat

1. Breeding resistant animals. Identical-twin studies in New Zealand have demonstrated a significant genetic effect on susceptibility to bloat although consistent differences between breeds, for example, have not been established. Not surprisingly, in view of the nature of its occurrence, there are few reports of estimates of the heritability of bloat resistance in cattle. The development of a simple test for susceptibility is important in the development of more resistant cattle.

2. Avoidance of risk. The elimination of the major offending legumes which form pastures and the encouragement of grass species is one way of avoiding the problem, although it is often costly because of the agronomic and nutritive features of the legumes which explain their predominance in suitable conditions. A more acceptable solution is the use of bloat-free varieties of the legumes involved. Some claims have been made for bloat-free clovers but the development is hampered by the lack of a simple effective assay method for testing the bloat-promoting properties of plant material.

3. *The administration of anti-bloat agents.* Several agents have been found effective in controlling bloat in grazing cattle at risk. The most effective of these is the twice-daily drenching of cattle with Pluronic agents. These materials have detergent properties which are thought to activate naturally occurring antifoaming lipids in the rumen. More traditional methods have involved the use of antifoaming agents such as paraffin oil or tallow, which reduce the surface tension of rumen fluid and collapse the tough-foaming properties. Antibiotics have also been used with some success, presumably stemming from their suppression of microbial activity and/or any effect on microbial flora.

Other methods not always so reliable in their use but nevertheless attractive to the farmer, because of their ease of administration, are:

4. *Flank application.* This involves daubing the cow's flank with 30–60 g of tallow or paraffin oil at milking times in the hope that the cow, by licking itself, will thereby regularly ingest sufficient antifoaming agent to prevent bloat.

5. *Incorporation in drinking water.* This involves the addition of about 50 ml of an agent like Pluronic L64 per 50 litres of drinking water. This method suffers, as do other agents incorporated in this way, from the variability in water intake in varying weather conditions.

Cattle may also take some time to become accustomed to the taste of materials in their drinking water.

6. *Spraying of pastures.* This method, if carried out rigorously,

offers a reliable method of bloat control. Cattle are strip grazed and the fence moved several times a day. It is recommended that tallow or paraffin oil is used and the recommended level is that calculated to result in an intake of about 75 g per day per cow. Areas should be sprayed no more than two or three days in advance of grazing and the treatment should be repeated after rain.

7. *The use of slow-release capsules.* Slow-release capsules have been developed which are lodged in the rumen. These have been shown to reduce the incidence of both severe and mild bloat but they are not yet developed to the stage where they can give the acceptable degree of control achieved through twice-daily dosing or pasture spraying.

Mineral and vitamin problems

In addition to the major feed-related metabolic disorders that have been discussed so far there are many others, discovered and undiscovered, that may at present be implicated in the unsolved cases that abound in field practice.

Major problems with known mineral and vitamin deficiencies, are avoided where the cattle are given a diet formulated to a specification of the kind shown in Table 5.4. In many other cases cattle are given concentrate supplements so that although part of the diet may be fairly ill defined in terms of nutritional status, most deficiencies can be avoided by suitable formulation of the supplement. These conditions usually apply to dairy cows and it is only beef herds and dairy heifer replacements that are normally subjected to long periods with no supplementary feeding.

Some of the main problems associated with such feeding occur in beef cattle and growing heifers during the grazing season and deserve some comment:

Copper

Copper deficiency can often be a problem in beef herds under certain circumstances. The cause may be a primary copper deficiency or an impairment of copper absorption and utilisation because of the interference of other elements such as molybdenum and sulphur.

Copper deficiency can cause a non-specific ill thrift in young growing calves and in cows that is difficult to confirm without analysing samples of blood serum or liver samples from casualties to assess the body status of copper.

More obvious symptoms include hair discoloration, very evident in black cattle which show typical rusty ginger or bleached greyish colour, sometimes as rings around the eyes. More extensive changes in the hair keratin affect the texture of the hair.

More prolonged copper deficiency can lead to bone abnormalities and lameness, to anaemia and in extreme cases to heart disease as evidenced by the 'falling disease' reported in Western Australia.

A common symptom, particularly associated with secondary molybdenum-induced copper deficiency, is severe scouring (diarrhoea).

Infertility in beef cows has sometimes been attributed to some of the effects of copper deficiency.

Predisposing factors. Copper deficiency is often associated with certain soil types particularly light sandy, peaty, limestone and clay soils and soil copper levels <6 ppm DM need to be looked at carefully.

Some pastures such as the 'teart' pastures in Somerset, U.K. have high levels of molybdenum and sulphur which render the copper unavailable and give rise to deficiency even where soil and pasture copper levels are not very low.

Copper deficiency may be accentuated by liming and by the presence of high levels of several other elements such as calcium, cadmium, zinc, iron and lead.

Diagnosis: Copper deficiency is best confirmed by assaying blood levels which for total blood normally range from 0.5–1.5 μg/ml or liver samples from casualties.

Treatment and prevention. Copper deficiency can be treated by injection of copper compounds such as copper edetate and by correction of the dietary deficiency (levels of 7–14 ppm of DM). Where molybdenum and sulphur levels are high, cattle need to be moved to safe pasture or regularly supplemented with copper. Any concentrate supplementation, particularly the creep feeding of beef calves, is an effective method of overcoming the problem. Pasture supplementation at the rate of 5–7 kg copper sulphate/hectare can also be effective when simple primary deficiency exists.

Cobalt

Cattle are able to synthesise vitamin B_{12} in the rumen provided the essential component, cobalt, is provided in the diet. Some soils are particularly low in cobalt (0.05 mg of cobalt per kg of herbage DM is required for cattle) and in these areas non-specific ill thrift is a common symptom. Cobalt-deficient soils ('pine' areas) are often ill-drained soils or soils subjected to heavy liming. Cattle may show signs of gradual progressive loss of appetite and condition even in conditions of adequate pasture availability.

The condition can be best confirmed by liver B_{12} assay levels although serum B_{12} may be a useful and more accessible indicator.

Treatment and control. Cattle in badly affected areas may justify the injection of vitamin B_{12} but since regular administration is required, the normal prevention is by incorporating cobalt in a feed supplement, dosing with cobalt sulphate at regular two-week intervals or using cobalt slow-release bullets which are administered by mouth and lodge in the rumen providing a small but constant supply.

Selenium/Vitamin E

The deficiency of selenium and vitamin E, or of both in combination, can give rise to muscular dystrophy in calves and is claimed to be implicated in some cases of retained placentae in cattle.

Certain soils and feedingstuffs have a low selenium status or a low vitamin E content.

Pastures with a high-sulphur content and high rainfall are sometimes associated with low-selenium status.

Other cases of interference with vitamin E are diets rich in unsaturated fatty acids as are some milk replacers; administration of rancid cod liver oil is a notorious example.

In such cases it is important to supply vitamin E and other antoxidants.

Storage of grain using propionic acid additives or in wet grain towers may also depress vitamin E content.

Treatment and control. Apart from ensuring adequacy of vitamin E and selenium in the diet it may be necessary to treat animals that have been affected or are at risk by injecting selenium or dosing with selenium salts. However care is necessary as selenium is easily toxic at excessive levels.

Other micronutrients which may be a problem under certain circumstances include iodine which can lead to low calf viability among other things and occasionally calcium and phosphorus. Deficiencies of vitamins A and D may also arise under certain conditions.

Finally mention must be made of the implication of nutrition in many other areas of cattle health. A well-known feature is the effect of parasites, which have a distinctly different fate according to whether the host is well nourished or otherwise.

REFERENCES

JULIEN, W. E., CONRAD, H. R., HIBBS, J. W. and CRIST, W. L. (1977), 'Milk fever in dairy cows, VIII. Effect of injected vitamin D_3 and calcium and phosphorus intake on incidence', *Journal of Dairy Science 60,* 431–6.

LENG, R. A. and MCWILLIAM, J. R. (1973), 'Bloat', *Reviews in Rural Science No. 1,* University of New England.

LIVESEY, C. T. and FLEMING, F. L. (1982), 'The importance of nutrition in the aetiology of laminitis and sole ulcer in Friesian cattle', University College of North Wales, *Departmental Note.*

MORRIS, C. A. (1991), 'Screening and selection for disease resistance—repercussions for genetic improvement', *Breeding for disease resistance in farm animals*, Editors Ascford, R. F. E. and Owen, J. B., C.A.B. International, Wallingford.

PAYNE, J. M. (1977), *Metabolic Diseases in Farm Animals,* Heinemann.

PATERSON, R. and CRICHTON. C. (1960), 'Grass staggers in large-scale dairying on grass', *Journal of the British Grassland Society 15,* 100–5.

PICKARD, D. W. (1981), 'Calcium requirements in relation to milk fever', *Recent Developments in Ruminant Nutrition,* Editors Haresign, W. and Cole, D. J. A., Butterworths.

WORDEN, A. M., SELLERS, K. C. and TRIBE, D. E. (1963), *Animal Health Production and Pasture,* Longmans.

Chapter 10

FUTURE DEVELOPMENTS

Development in agricultural techniques combines the painstaking process of refining existing systems with that of the incorporation of unpredictable new ideas which often seem to conflict with accepted dogma. Exploring future possibilities can guide the original, inquisitive researcher/practitioner into fruitful avenues whose outcome can only be fully realised after they are tried out in practice.

FUTURE DEVELOPMENTS in cattle feeding can affect both the feed materials used and the way they are utilised.

FEED SOURCES

In looking at the new feeds that could become available there are several possibilities both in supplying the roughage and the concentrate part of the diet.

New Forage

Already there are several dairying situations where a scarcity of roughages has led to strong pressure to explore new sources. In arid and semi-arid areas the production of forage crops under irrigation is expensive and sets a limiting factor to the development of dairying. In such cases dairy cow diets based on cereal straw and other by-products of dry land crops may be a more suitable alternative. Such materials can be cheaply chopped so that they may be incorporated with concentrates into complete diets. Since the optimum proportion of chopped straw in the diets of high-producing cows lies in the range 20–30 per cent it is conceivable that whole-crop harvesting can be employed to provide a mixture of straw and cereal which can be suitably supplemented to bring the whole mix into the required composition.

By-products of several industries may be considered as possible bases for cattle diets. These include wastepaper and other wood products, provided they are free from toxic contaminants. There is already sufficient knowledge about certain wastepaper materials to suggest that they could be viable cattle feed sources.

Another possible source of high-fibre material is livestock solid waste. Poultry droppings have already been used, primarily as a protein source and recent indications are that pig dung can be used as a forage replacement. One of the drawbacks of the use of such material is that, if it is to be widely available, it needs to be dried and this of course markedly increases the cost of the product. Also stringent steps need to be taken to ensure that there are no risks of disease transmission.

New Concentrate Sources

The process of exploring all possible crops for cattle feeding purposes goes on and with the increase of communication between different parts of the world many areas will become more familiar with products grown elsewhere. For example, many British compound feeds have recently incorporated cheap materials like manioc instead of the more traditional cereal grains.

The emphasis on the role of undegradable protein in cattle feeding gives greater impetus to the incorporation of processed animal byproducts as part of cattle diets, for example fishmeal and meat meal. However the health risks, particularly in relation to the transmission of diseases like BSE, are likely to prove a deterrent to this practice.

Biotechnology may eventually yield results of great value to the cattle feeding industry. Applications are being explored where selected microbes are grown on suitable media to give a high yield of microbial protein. This can be high-quality material suitable for use in the diets of calves and lactating cows. Since new strains of microbes suitable for a wide range of media may soon be developed, this avenue is a highly promising one. Whilst people will take a long time to become accustomed to microbial products in their diet, it is not a new departure for the cow since she already obtains much of her nutriment in this way. It is likely that the ruminant will soon be beaten at her own game by huge artificial fermentation vats since these provide better control over conditions and thus facilitates the manipulation of bacterial strains.

Additives

Already a substantial array of additives are used in cattle diets; these include known vitamins and minerals as well as growth- and lactation-promoting materials. It is unlikely that all the micronutrients of importance to cattle have already been discovered and as increasingly sophisticated analytical techniques become more widely used it is probable that additional requirements will become known. The field of infertility in cattle and its association with the diet is a morass of uncertainties and complexities, where only some of the major uncomplicated effects have so far been unravelled. It is likely that many of the known micronutrients interact together in their effect on the cattle beast.

In dairy production the use of lactogenic additives went into abeyance since the work in the earlier part of this century when the use of iodinated casein and similar thyroid-promoting substances were examined. If, however, high-producing Holstein-type cattle and three-times-a-day milking have an economic role, why not lactogenic additives? In the one case an attempt is being made to increase the cow's lactogenic prowess through breeding, a process that proceeds at little more than 1–1½ per cent per annum at best. In the other case extra stimulation is used as a means of increasing lactogenesis in cows of varying production level.

Additives which could be selectively used to promote lactation in medium-potential cows could help partition feed into milk rather than body condition without over-straining the cows that already produce high yields of milk.

Discoveries such as that of the crucial role of undegradable protein during early lactation in promoting partition of energy intake into milk may be extended to other micro-nutrients which may be added to the diet in small amounts to achieve the same end result.

Interest in this field of lactogenic substances has been opened up through the work on growth hormone. The use of bovine somatotropin given as daily injections to dairy cows has resulted in yield increases of 20 per cent and more. It is too early yet to decide whether improved methods of administration to cows and acceptance by the consumer will make this an attractive proposition. As with other lactogenic additives, its value may be more for stimulating lactation in the lower yielding fraction of a herd than in further stimulating good cows at their peak.

FEEDING SYSTEMS

In the recent history of developing feeding systems and allowances for cattle there has been a tendency to get away from the menial feeding trial to an attempt to predict what happens in the system as a whole from the behaviour of the detailed components. Whilst this approach, as embodied in the ME system (MAFF Bulletin 33; ARC 'Nutrient requirement for ruminants') has its place, it has not led to the most useful guides to the practising feeder. There is a crying need for more simple but well-conducted trials, such as those carried out by Gordon in Northern Ireland which can give data to confirm and refine the practical input/output information upon which tables, like those in Chapter 7, are based.

Many cattle feeding systems in a particular region or country are relatively restricted in their form and in the range of ingredients that are used, for example, the type of forage grown.

This means that feeding trials for beef and dairy cattle based on a fairly narrowly defined representative system could give *real* data on input/output which would have a sound validity for general application in that they take into account the associative effects of mixed feeds. Being a simple approach which includes the realities of product prices and input costs, and one that also takes into account measured realised intakes, it is far more likely to be used by practical feeders.

Maximum Profit Techniques

On the sound basis provided by such practical data the full power of computing techniques can be employed to extend diet formulation into the wider realm of farm management decision-making, the ideal diet for a dairy herd being the one which helps maximise the whole-farm profit.

Automation in Cattle Feeding

The cattle industry is poised on the brink of major developments in automation. The question is to what extent these influences will affect feeding practice. The development of programmed electronic feeding for the individual is now a real practical possibility and some development has already taken place. It is ironic that just when all was set for a real boom in this respect, the advantages of individual allocation of feed according to production stage have come into question. On present evidence simple methods of flat-rate and complete-diet feeding appear to be more appropriate than the expense of striving for precise individual allocation. In smaller units the simple system of self-fed silage in conjunction with a standard concentrate allocation, dispensed in three or four meals is still viable.

On the larger unit the flexibility of using a mixer waggon and adopting a simple complete-diet system appears to have many advantages which are capable of further exploitation.

The development of automatic milking could allow a complete change in attitudes towards frequency of milking. Undoubtedly such a far-reaching development will have repercussions on feeding practice which, as yet, can be perceived only dimly.

Storage Feeding

In temperate areas the length of the grazing season has been determined from the overall assumption that, within the constraints imposed by grass growth and ground conditions, pasture use should be maximised. This situation is now changing and the extreme of keeping cattle entirely on conserved grass is becoming a practical choice. The feeder then is confronted by a question of what optimum to choose within the full range from zero to maximum grazing. Storage feeding of beef cattle in yards can be fairly simply evaluated from known production parameters including all the normal technical yardsticks. Such an example is shown in Table 10.1 comparing a conventional with a full storage feeding system.

Table 10.1. Storage feeding of dairy cows; preliminary results from Crichton Royal, Scotland

	High-concentrate group	*Low-concentrate group*	*Top 25% of dairy farms in England and Wales*
No. of cows	24	24	
Milk sold per cow (kg)	6,165	5,417	5,743
Concentrates per cow (tonnes)	2.22	1.37	1.87
Silage DM intake per cow (tonnes)	3.47	3.73	
Cows per forage hectare	3.34	3.09	2.41
Margin over purchased feed (£/cow)	584	577	553
Margin over purchased feed (£/ha)	1,951	1,783	1,333
Milk fat (%)	4.16	4.05	
Milk SNF (%)	8.95	8.82	

Source: Ley, Coiin (1983), *Farming News* 22/4/83, M.M.B. (1982), *Analysis of FMS costed farms 1981–82.*

In many circumstances and in many parts of the world the storage feed system is becoming the more profitable. In dry areas such as the mid-West of the USA, where grain growing abuts on to cattle rearing ranges, feedlots based on storage feed have been a viable business proposition for many decades. There seems no reason why existing cattle accommodation in wetter areas of the world cannot be used for twelve months rather than four to seven months of the year if silage clamps can be made available.

For dairy cattle the main constraint to the adoption of storage feeding in the wetter areas of the world is the present unsatisfactory state of normal cow accommodation. many cattle units use the cubicle system which affords great savings on bedding material but is not entirely satisfactory for all cows. Many dairy farmers are now only too glad to get their cows on to grass in the spring in order to get them clean and sound on their feet; the prospect of keeping cows under normal cubicle conditions all the year round is not an inviting one. If this problem could be overcome, possibly by using sacrifice paddocks for summer idling areas, storage feeding of dairy cows in temperate areas could become a viable proposition. In conjunction with three-times-daily or automatic self choice milking it would make an attractive combination in certain circumstances.

In between the two extremes lies the exploration of less extreme combinations such as that of delaying turnout until after an early cut of silage from the grazing area, or of feeding the animals indoors overnight and grazing them only during the day. Another possibility is the earlier cessation of grazing and utilising the pasture thus freed for other stock, such as sheep or store beef cattle, that are not so sensitive to the vagaries of autumn grass supply as the high-producing dairy herd. Already the intermediate system called "buffer feeding" is gaining popularity. This entails supplementing cows during the grazing season with conserved forage or complete diets. This system provides an insurance or buffer against the vagaries of seasonal grass availability and can allow more aggressive stocking rate policies since the fear of depressing cow yield, through pasture shortage, is removed. Buffer feeding allied to the maintenance of an optimum sward height could be the optimum answer in wetter grass growing areas, whilst complete storage feeding with access to exercise paddocks could be more suitable in drier areas (Phillips, 1989).

The stage has been reached where this situation can be modelled sufficiently well on the computer that these combinations can be explored. As and when promising combinations are revealed from computer modelling output, these could be tried out in practice on development farms to see whether the whole adds up in practice. Feedback from such development exercises could help refine the basis on which the modelling rests.

REFERENCES

PHILLIPS, C. J. C. (1989), 'New techniques in the nutrition of grazing cattle', *New techniques in cattle production*, Editor Phillips, C. J. C., Butterworths, London, p. 106–20.

Index

Farming Press Books

Listed below are a number of the agricultural and veterinary books published by Farming Press. For more information or a free illustrated book list please contact:

Farming Press Books, 4 Friars Courtyard
30–32 Princes Street, Ipswich IP1 1RJ, United Kingdom
Telephone (0473) 241122

The Principles of Dairy Farming • KEN SLATER

An introduction, setting the husbandry and management techniques of dairy farming in its industry context.

A Veterinary Book for Dairy Farmers • ROGER BLOWEY

Deals with the full range of cattle and calf ailments, with the emphasis on preventive medicine.

Cattle Ailments—Recognition and Treatment • EDDIE STRAITON

An ideal quick reference with 300 photographs and a concise, action-oriented text.

Calving the Cow and Care of the Calf • EDDIE STRAITON

A highly illustrated manual offering practical, commonsense guidance.

The Herdsman's Book • MALCOLM STANSFIELD

The techniques and skills of cattle husbandry for students and farmers.

Calf Rearing • THICKETT, MITCHELL, HALLOWS

Covers the housed rearing of calves to twelve weeks, reflecting modern experience in a wide variety of situations.

Farming Press Books is part of the Morgan-Grampian Farming Press group which publishes a range of farming magazines: *Arable Farming, Dairy Farmer, Farming News, Livestock Farming, Pig Farming, What's New in Farming*. For a specimen copy of any of these please contact the address above.